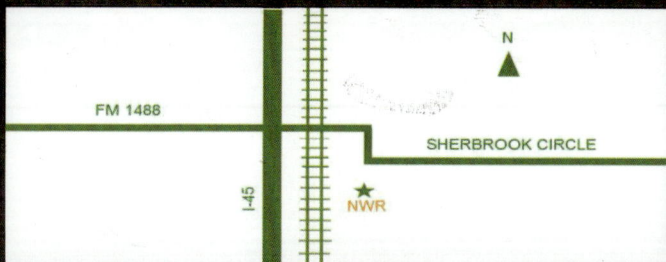

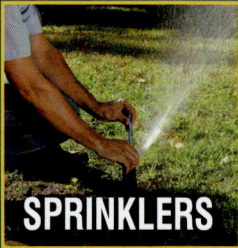

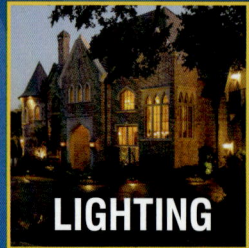

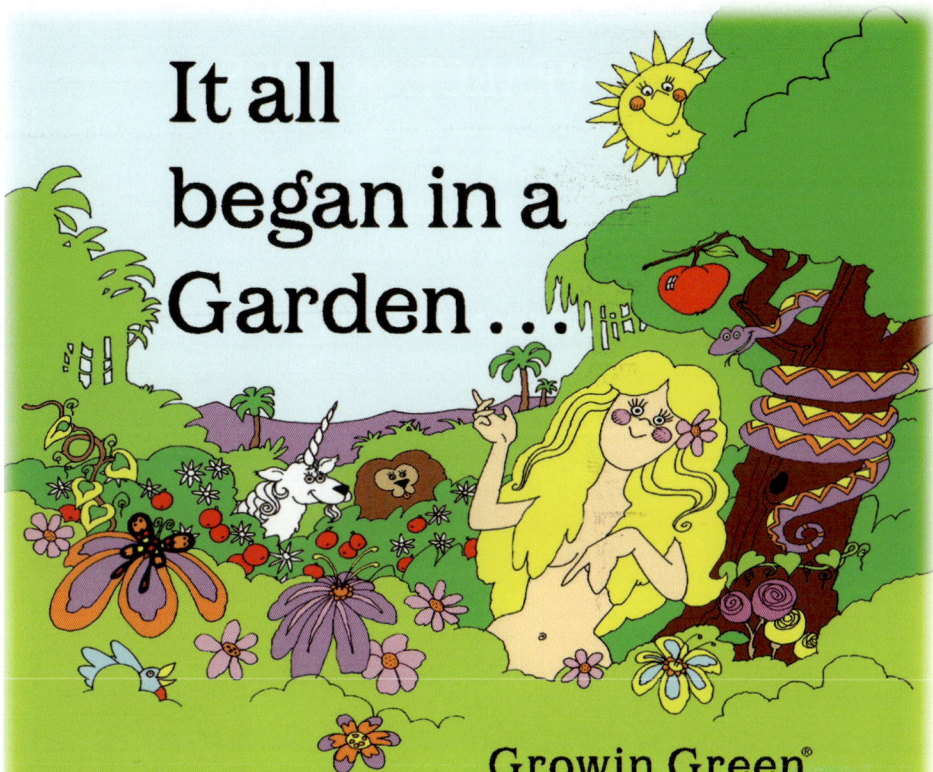

New Decade Gardening
A Guide for the Gulf Coast

Front & Back Cover Photos by

Robert Spangle Photography

Robert@RobertSpangle.com

Editing by

Yvonne Salce Lemmon

New Decade Gardening

Introduction

INTRODUCTION

When I set forth to write this book, my initial intent was to reproduce my 'first ever' self-published book known as *Gulf Coast Gardening with Randy Lemmon*. *Gulf Coast Gardening* wasn't technically my first book, but it was my first self-published book. It was published 15 years ago, and boy did that prove to be a bit outdated to say the least. The basic concepts and chapters made for a really sound "rule book" for gardening along the gulf coast, with a major emphasis on landscaping and lawn care
.

However, with so much outdated information, there was obviously a need to rewrite the book with solid advice to carry the home gardener through 2020 and beyond. In fact, my gut instinct was to title the book *Gulf Coast Gardening 2020*. But my wife (a superior writer, whose advice I trust) and a few others pointed out that people would probably not purchase that book in 2021 and beyond. Instead we settled on *New Decade Gardening*.

Some people may already be asking, why should I buy this book, if it's just a rehash of his first self-published book? Well, let's point out that it's almost been 16 years since *Gulf Coast Gardening* with Randy Lemmon was produced, and if you know how things change every couple of years in the world of horticulture in the first place, you'll understand that this is absolutely the latest information you are going to need. Secondly, as I was reminded before my first book ever went to print, (and this is a quote from a co-worker who served as a producer of GardenLine for years, and helped proof that book) "This is the ultimate rule book for gardening along the gulf coast, and it's like having GardenLine at your disposal every day of the year."

Over the years, I've also had a lot of comments and questions about why there aren't pictures in the book. The answer is that while many gardening books are chock full of pictures, they are seldom filled with any practical advice. I could provide beautiful pictures too, but I think you are better served if you have a complete understanding of all the concepts that make gulf coast landscapes work. The GardenLine radio show is all about questions and answers, and we are quite successful in that vein. We are also fortunate in today's technology of social media, that we can still answer questions, and do it with the help of pictures on social media. If a book of this size is loaded with pictures, it raises the price significantly, and can take upwards of 3 years to produce. I like the idea of keeping the costs way down, and affording you the opportunity to get your hands on the book much quicker.

I also know this book is exceptionally helpful for anyone who is new to gulf coast gardening. I make the joke on the radio show all the time to people who have just moved here from other northern and western states, that they now have to forget everything they ever learned about gardening and start over. A concept that may best be summed up in an email I got 16 years ago, right before I self-published *Gulf Coast Gardening*.

Randy, Thank you for saving my trees from the dreaded Crape Myrtle Massacre. We moved here in the early 2000s from Michigan. At final inspection we were told "Oh, you cut this to about 3 feet every year around February." So we moved in at the end of February and I cut my crape myrtles all the way back...Then I started listening to you... just 3 years later, while everyone else's Crapes are ugly, my Crapes are the envy of the neighborhood. – Debbie J.

Chapter 1

Building the Perfect Beds

– It's All About the Soil

I've been reminding people for over two decades that, "All gardening success in Texas starts with the soil." Somewhat sadly, we have to make those soils happen, because they don't always occur naturally. This message really needs to sink in with folks who have moved to this region from northern states. Up there, while you can't garden for 5-6 months out of the year, when you can you're able to just throw things into the existing soil and everything works.

We simply can't do that here, because of the clay soil that abounds in the region. Some call it clay, others call it caliche and some call it gumbo. Whatever you call it, it simply cannot sustain much plant life without being amended like nobody's business with organic matter. I distinctly remember feeling so clever when I wrote my first book about gardening, and the soils chapter was titled "Planting in SE Texas Gumbo – Not a Cookbook Chapter!" I learned that people passed it over, and didn't thoroughly read it, because it came across to newcomers as a joke.

This time around, I need you to take this chapter very seriously. For the die-hard fans, or someone who has all my previous books, you will notice right away, that this chapter is a compilation of everything I've written about soils and mulches for the past 15 years. It will focus almost exclusively on "building beds" for landscapes and vegetable gardens, with a tiny emphasis on "top dressing" for the turf. Towards the latter half, I will get even more serious about mulch. We'll talk about the good, the bad and the ugly – of course that's the dyed mulch.

It's interesting to note in the evolution of building beds for this region that rose soil is still the best starting point. Before I became a garden guy on Houston radio, my predecessors and my agricultural mentors at Texas A&M University talked about making your own rose soil. And it was pretty sound and simple advice: Mix together equal parts loam(y) soil, sharp sand and organic matter (preferably compost, humus etc.). But honestly there weren't many soil yards around where you could get those ingredients in bulk to blend your own.

Now, just about everyone makes a rose soil or azalea soil. There are soil yards that have "landscapers mixes" as well as "vegetable garden" mixes. You name the specifically needed planting medium these days and someone makes it in bulk and eventually in bags. So, since we've come a long way in the development of soils, it still baffles me why anyone would try to plant landscape material or vegetable gardens in our existing soil. I'm often equally mystified why people use peat moss or plain old potting soil in outdoor beds and landscapes.

Avoid Peat Moss in Landscaping

Before we get to the rules of building beds and amending soils for landscaping and vegetable gardening purposes, let me harp on the peat moss/sphagnum peat issue, and why it should never be used here in gulf coast gardening. These "organic" garden soils or "organic" raised bed soils derived of sphagnum peat moss and processed forest products have a GREAT water holding capacity - like a sponge. The job of peat and sphagnum is to hold moisture in the soil (where the plant roots are) and when the soil stays wet and cannot dry out, plants cannot produce food/sugars. They are essentially drowning. The other polar opposite effect, and negative for horticulture too, is when peat-heavy landscape soils dry out too much. This is a real problem when people mistakenly use peat-heavy bags as a mulch. The dried out peat will create an impenetrable barrier at the top, casting/shedding off any kind of water. Thus, nothing percolates down to the soils. I see this so often in my consulting business.

And there is the "environmental" slap in the face. As other gardening experts have been known to say: "...the biggest problem with peat moss is that it's environmentally bankrupt!" It is essentially "mined" by scraping off a bog's top layer of living sphagnum moss. Despite manufacturers' claims that bogs are easy to restore, the delicate community of animals and insects that inhabit a bog cannot be quickly re-established. Yes, peat moss is a renewable resource, but it can take hundreds to thousands of years to form." Peat moss is the partially decomposed remains of sphagnum moss, harvested from bogs.

Here's my simple rule: If the first two or three ingredients listed on a bag of soil are peat moss or sphagnum peat moss, avoid it at all costs for landscaping. You and I alone may not be able to change the environmental degradation resulting from peat bog harvesting, but if more of us use peat-free products, like those I endorse, maybe the message will be heard.

Peat is not the amendment we need to use in clay soils; we need to add more compost and shale to improve our clay soils. The rock will open up the airways for the plant to breathe.

We cannot see soil air spaces, just as we cannot see our lungs. We can only see whether plants do well or not. If your plant hasn't grown in 2 years, which automatically tells me, there are no air spaces for the roots to grow so the plant just sits and normally declines. As we are gardening in clay soils or small particle sands, every time we put a shovel in the ground, we need to add compost and expanded shale to create more air spaces in the heavy clay soil. You only get one shot to get the planting right! My advice for years was to start with quality Rose Soil at the very least, which does have humus/compost in it, but the ones I recommend have never had any kind of peat.

Okay, so here's pretty much the message I've had for years about building raised beds for LANDSCAPING PURPOSES. The protocols for VEGETABLE BEDS is coming soon after this.

Building The Raised Bed - Landscapes

I can make this very simple. You buy rose soil, build it up 10-12 inches and lock it in with some kind of stone, rock or landscape timber. Thanks! Drive safely, and we'll see you again real soon! Okay, it's a bit more detailed than that, and there are nuances along the way.

Let me start with the basic technique - tilling a few inches of rose soil into the existing soil, then building the rest of the bed on top of that. While it's not necessary when transplanting one-gallon or smaller plants, some come in 3, 5 and 15-gallon containers, and 10-12 inches of rose soil is simply not going to be enough depth for the 15-gallon sizes and larger. So, you'll have to dig into the clay soil. If you till in several inches of rose soil with the existing clay, a few inches of larger transplants' root zones will benefit from the mixed-in organic matter, and the plant will have a fighting chance of success.

Think about it this way: If I dig a hole in existing clay soil and slide a plant into it - like a shotgun shell into a gun - what do you think is going to happen to the root system as it tries to grow? And boy, do they want to grow. Let me answer that. The roots will wrap around each other and eventually just quit because there's no place to go. But, imagine a few inches of clay blended with rose soil or organic matter, then the raised bed material on top of that. A 10-gallon small shrub's roots are just over 10 inches deep in the container. With the amended area down low and rich, friable soil in the top 8 or so inches, you'll see how building raised beds is the key to landscaping success in our area.

Pretty much each weekend on my radio show, I hear stories from newcomers to gulf coast gardening like this: "Randy, I planted a knockout rose last spring, just after moving here from (fill in a state north of the Mason Dixon line), and it hasn't grown much at all. But my neighbor's rose is flourishing." I'll usually ask how it was planted, and … you guessed it … they dug a hole in the existing soil and slid the plant right in that clay tube. Have you ever looked at the delicate roots of something just slid from a container? How in the world could those feather-light roots penetrate that clay wall?

What About Existing Beds?

Over the years, many have asked me if they can convert existing beds to raised beds. The answer is complicated. It's not just a YES or NO! So, here are some basic rules if you want to try a transformation.

First, you'll have to extract all your existing smaller shrubs and landscape elements and start over using the bed-building tenets just listed. Second, you'll need to aerate the soil, occasionally adding good mulch, soil activators and micronutrients. You can also accomplish this over time by updating mulch levels with organic products like compost or native products at least twice a year, and poking holes occasionally with something like a piece of steel rebar or a metal rod. I also encourage spraying or even soaking in something like a soil activator or liquid organic foods a few times a year.

You can leave the larger shrubs in place and build raised beds adjacent to them, where smaller new plants and flowers have a better chance at succeeding. If the larger shrubs and trees just aren't growing, however, re-set them and any small trees using the process we talk about in the Trees chapter.

If you haven't read this whole book yet, know that the soils and technique needed for trees is in the Trees chapter: It's known as the *"Twice as Wide and Half Again as Deep"* method.

Soils for Lawns and Turf

I'm amazed at how turf "top dressing" has evolved from the use of bank sand to compost in the past 20 years. Up until 1996 in Houston, you top dressed a yard with sand or loam. Then, along came soil scientists who developed high-quality, highly-screened composts that added the tilth that was long sought after. And along the way, they discovered that it solved many of our problems with diseases and weeds.

Compost top dressing should be the norm now. And thankfully there are bulk soil companies and others that bag up high-quality composts for southeast Texas. There are also some enriched soils that can be used as top dressings, too. Just make sure it's enriched with high levels of organic matter, along with the loam soil. Sadly, I have seen an "enriched soil" used as a top dressing that was obviously just clay-based soil, devoid of any kind of organic matter. And a mistake of that magnitude, can actually kill a yard.

If you're buying or building a new home, I think preparing the soil before you lay sod or sow seed is the best investment anyone can make. You will be ensuring a healthy environment in which root systems can set up shop. If you've ever driven around a new subdivision, you've probably seen new yards going in on top of scraped-up, clay soils. How are those roots going to establish? So, If you can get in before sod is laid and prep the soil with compost, or even just a few inches of permanent soil amendment, your turf will lock in its roots much quicker and much deeper and ultimately give you the healthiest lawn on the block. Unless your neighbor also has this book.

Building the Perfect Beds – Vegetable Gardens

I can have fun with a simple answer here as well. Mix two parts rose soil to one part quality compost, and make a raised bed of 10-12 inches. Boom! That's all you need to know!!

Yeah, I know it's not always that simple. But it can be. I'm not here to argue with any commercial growers about the need for raised beds in vegetable gardening. I completely understand that they incorporate sand and organic matter in the fields they have worked and amended for years. I have to form my advice for typical homeowners in the suburbs of southeast Texas. And that means, we build raised beds, plain and simple. And when we experience warmer-than-normal winters, we also need to get busy with building vegetable beds as early as possible.

For years on GardenLine and in my books, I've tried to make it as easy as this: Make a raised bed of good garden soil — equal thirds of soil, sand and humus (rose soil) — as the starting point. It should be at least 6-8 inches, but 10-12 inches will be even better. Then, amend that with well-composted organic matter, humus or manure at about one inch (tilled in later) to every 4-6 inches of good soil. These days, though, there's an even simpler recipe: Two parts rose soil to one part compost. Then, lock it in somehow. Use timbers, lumber, cinder blocks or landscape stone. Just lock it in!

Most people I know (myself included) who use this technique don't even try to kill the grass or weeds in the area. You can suppress grass or weeds with 8-12 layers of newspaper, covered by the soils and composts. Over time, the paper will break down and become part of the soil. In the meantime, it prevents weeds and grass from growing up through your pristine vegetable garden. Best inside advice when doing this, is to wet the newspapers so they stick to the ground below, and when you start layering the dirt they don't move.

You don't have to cover the top layer with mulch, unless you want to suppress weeds and if you want to conserve moisture. But if you're like me, I simply use compost as my mulch these days, and that's slowly and surely adding more organic matter to the raised bed over time. There's a lot more to learn about mulch in general right here.

Much Ado About Mulch

I could not walk away from this soils chapter without addressing the best mulching habits for topping the soils discussed above. Anyone who has purchased my other books or follows me on social media is already wise to my disdain for dyed mulch. Beneficial mulch is absolutely needed for landscaping in this region, but we have to end the dyed mulch debacle. If just one in 100 people who read this book will stop using dyed mulch, I feel that writing the entire thing will have been worth it.

So, I'm here to set the record straight on all things mulch with my Top 10 Commandments of Mulching as well as the best choices.

First, though, let's get some basic science on the table so you know I don't just have a personal vendetta against dyed mulch. My argument is based in soil science and plant health. For the record, there are scores of colored mulches that have been significantly improved through the incorporation of organic matter and the use of organic dyes. But I have yet to find one that I would endorse.

In any case, it's the overtly black-dyed mulch that is the worst offender, because its mostly made from chipped up wooden pallets. And it's not just my opinion. Dr. Harry Hoitink, revered soil scientist and professor emeritus at The Ohio State University says "Wood mulches can slow the growth of established plants and just plain starve new ones to death by 'tying up' the available food in your soil, a process known as 'nitrogen immobilization." He's one of the nation's leading experts on mulch, and he goes on to say, "Wood is carbon, and carbon always looks for nitrogen to bond with so it can break down into new soil."

Think about that for a second, and you'll understand that's the principle behind composting. Wood mulches take nitrogen right out of the soil, out-competing your nitrogen-needy plants. And when you take the wood in these old pallets and spray it with dye, you've got the worst type of mulch for use around new landscapes, especially on shrubbery, annuals, perennials and smaller plants.

Now, understand some differences. The double-shredded hardwood mulches I recommend are composed of finely shredded wood, mixed with compost. Dyed mulches are nothing more than chips and chunks of wood — neither mixed with anything organic nor shredded. Also, the dye will obviously leach into the soil.

So, why are there loads of landscapers who don't know this? I surmised long ago, they are just flat out uneducated on the subject. Some landscapers do actually know this stuff but are just too lazy to explain to customers why dyed mulches should never be used. A homeowner says they want black, and rather than run the risk of losing business, the landscaper gives it to them. There also may be some shady enough to know the deadly mulch will sicken and kill the customer's plants. So they'll pick up some future replanting business. And there are the unprincipled landscapers who know that black-dyed mulch loses its color quickly, so they can count on constant reapplications throughout the year. Finally, there are some landscapers who are just out to lunch. They think it looks good! I'm sorry, but I don't think it looks classy at all. I say it looks unnatural.

I think the black-dyed mulch phenomenon happened because it was simply a great way to recycle wood debris and numerous pallets. Someone got the idea to shred it and make it a "mulch" by dying it. They never thought to study the science and biology behind its use. But it became a major money-maker, and it took off. Couple that with so many transplants coming from states where they never use mulch and the soils are naturally dark, dark, dark! Maybe that need to achieve something that looked like home helped create this bizarre market. And if you're a real estate agent, and you encourage people to use black-dyed mulch to stage a house, shame on you too! Tell them to get a fresh application of Texas native mulch for curb appeal.

So now that you know better, please cease the use of black-dyed mulch forever. And remember that the mulches I've recommend are Texas-native varieties that also include ample compost - beneficial to the soil, not harmful.

Here's my Top 10 Commandments when purchasing and applying new mulch, followed by a list of preferred mulches.

I - Thou must understand that wood chips don't equal mulch. - Mulch isn't shredded wood. Mulch covers the soil to retain moisture, provides nutrients and prevents weeds. Wood chips or shredded wooden pallets don't do this. They actually rob the soil of nutrients as they break down.

II - Thou must never use wood shavings as mulch. - Too often, I'm asked about using wood shavings or sawdust from a freshly cut tree, woodwork projects or a ground-out stump. Fresh wood, as mentioned above, starves plants. Wood is high in carbon, and carbon seeks out nitrogen as it breaks down into the soil. Fresh wood shavings will immediately deplete the area of nitrogen, and the plants will start turning yellow. Say it with me - nitrogen immobilization.

III - Thou must keep mulch away from foundations. - It's extremely important to keep mulch several inches below the point where the house and the foundation come together. Otherwise, insects or moisture can use the mulch as a conduit to weep holes and enter the house.

IV - Thou shalt not believe charlatans who claim mulch attracts termites. - Yes, termites might use mulch near a weep hole as a pathway in, but just having mulch doesn't attract termites. If you use the kinds of mulch I recommend below, you'll never have to worry about this, because termites actually need big chunks of wood on which to feast. This is a crock perpetuated for years by rubber-mulch purveyors and shady pest-control operators.

V - Thou shalt not make mulch "volcanoes" around trees. – You see these especially in subdivisions being marketed by builders. Mulches of all kinds are piled up nearly two feet tall around the base of a tree. My head just about explodes when I see this practice because it's incredibly unhealthy for the tree. Mulch rings around trees are good, if you use the right kind. But it only needs to be several inches tall at best. And while I'm on the subject, stop planting flowers in mulch rings.

VI - Thou must ask one's self, "Would I let my kids play in this?" - This is sort of a trick commandment. While I wouldn't recommend wood mulch with no compost for landscapes, I think they're good in playgrounds and on running trails. But as for dyed mulches, just stick your hands in a batch and rub them around. Then, look at your hands and tell me if the residue doesn't look like ashen soot or dye. Really, would you let your kids play in that? I know kids don't usually play in flower and landscape beds, but look what it does to your gloves, hands and clothes when you spread it out!

VII - Thou shalt not be fooled by "good deals." - You get what you pay for! Good mulch is seldom less than $3 a bag. It's always more cost effective to buy in bulk quantities. But you'll see tons of signs and ads that say "Mulch: Five Bags for $12!" or "Five Bags for $10!!" That is almost always wood-chipped mulch or dyed wood-chipped mulch. And nothing good can come from that.

VIII - Thou must avoid dyed (unnaturally colored) mulch. - Most dyed mulches are made from questionable wood supplies, like shredded pallets, and nothing else. Plus, no dye is good for the soil, plain and simple ... even if it is supposedly organic. It's still dye, and the dye will leach into the soil. Other ash-infused mulches are just plain caustic.

IX - Thou must understand that rubber mulch is the worst. - Bet you didn't think I could slam anything harder than dyed mulch, did you? But let's defend the use or rubber mulch in a couple areas. I can see the need to recycle old tires. I just don't want them in my landscape. I have seen rubber mulch used successfully in dog runs, hiking trails and kids' playgrounds. However, in the landscape, what leaches from rubber mulch is 10 times worse for the soils than that which leaches from dyed or ash-infused mulches. It introduces dangerous levels of zinc and other harmful chemicals that can kill root systems. Plus it heats up unmercifully in our summers, also killing roots.

X - Thou must believe there is no better mulch than compost. - Dyed mulch has become so dang popular along the Gulf Coast because them-there Yankees done moved here from other states. Seriously, people from northern states like Ohio and Pennsylvania are accustomed to darker soils than ours, and I think this may have been their attempt to achieve that look from back home. In truth, those northern soils are not really black, just really dark brown. Meanwhile, you can get that dark color by using really good compost as a mulch. Plus, I love the idea of good compost as a mulch mainly because it's feeding the soil, feeding the roots and feeding the plants.

So, what mulches can you use? Here are my top five:

Compost - By far the standard in Southeast Texas, and three are from the best soil purveyors around: Nature's Way Resources, Heirloom Soils (Warren's Southern Gardens) and The Ground Up. Their composts are dark, and exceptionally beneficial to the overall health of the soil.

Shredded and double-shredded hardwood mulches – Look for anything labeled "Native" or "Texas Native." The really well-shredded varieties will have a bit of compost in them as well. They come in bags, but rarely will you find the "native" varieties at mass merchandisers. The best deals are always going to come in bulk. The best bags of compost are found at independent nurseries, garden centers, feed stores and small hardware stores.

Mixed mulch (blended) - These mixes of shredded hardwood and shredded pine bark mulch give you a darker color that lasts longer than plain old shredded mulch. However, it won't hold that dark color near as long as compost. One of my favorite blended mulches these days is from Landscapers Pride, and they have a mixed mulch known as Black Velvet. I say don't let the name fool you, but know that this is that ultimate blend of compost, pine bark and shredded hardwood mulches that keeps it's dark color longer than any other.

Shredded pine bark – I would still rather have it blended with the shredded hardwood mulches above. But if this is the only mulch you can find, I can give tacit approval. Just don't make it the mulch you use all the time.

Pine straw – The Rodney Dangerfield in the gardening world since it just doesn't get the respect it deserves. There's usually plenty of it around, and it should be used more often, especially around evergreen shrub beds with plants such as azaleas, gardenias and camellias. But it comes with an aesthetics issue in that you have to like the needles' rusty-orange look.

Avoid Mulch Volcanoes

Whether you've read the Trees Chapter yet or not, there's a description in there about how right after planting the trees, you kind of 'berm up' a ring of soil around the tree upon its initial planting so you can focus the much-needed water to that newly planted root ball.

It is that 'berm' that has unknowingly and quite innocently been the reason mulch volcanoes have crept into our landscaping world and unfortunately led to this onslaught of tall mulch mounds around the trees. You see, the berm ends up being the template for the average homeowner for where and how far out to mulch trees, when actually you should only be putting down a very thin layer of mulch. In fact, the more established a tree gets, the more you should be opting to show off the Root Flares – the big roots coming right off the base of the tree, and looking like arbor anchors in my opinion.

So, a homeowner who doesn't know any better piles up a few inches of mulch inside the berm and they keep adding to it and adding to it, and the next thing you know they have a foot high mulch ring up against the trunk of the tree. I see this in my consulting business regularly.

If you're unfamiliar with the phrase "mulch volcanoes," these are the mounds or cones of mulch, one, two or three feet deep, piled up around the trunks of trees and shrubs. I often say I am not sure how this practice developed, but I noted earlier the "berm" is probably the first culprit. I also suspect uneducated and untrained landscaping companies (can't really call them professionals if they do this, right?!) are the originators. And it seems to be especially pronounced at model homes in new residential communities around the region. Then, you can easily assume that naive homeowners (or first time homeowners) spot the practice especially on model homes and decide to copy. Then it's simply a domino-effect in a subdivision. One neighbor sees another one doing it, and/or they all employ the same landscaper/mulch provider who is completely ignorant to the negative effects. And trust me when I say the negative effects will eventually kill the tree.

Professional landscapers get paid to do the work their customers order, and when homeowners ask for fresh mulch a couple of times a year ... as they should ... several inches sometimes get piled up in an area that might just need one inch. As it accumulates, it starts to cover the bark of the tree trunk and the very important root flare. That bottom section needs air and light. With excess mulch, it's forced into darkness and subjected to moisture. Tree bark that's too moist for too long will rot! And rotted bark cannot protect the tree from diseases. In fact, diseases grow better in the dark moisture of the mulch, and then the trees become susceptible to insects and diseases.

Some trees, such as maples, have shallow roots. If bark mulch is piled high around their trunks, the roots will start to grow into the mulch. These roots tend to stay in the mulch where they grow to encircle the trunk. This is called a "girdling root," and as it grows in diameter, it pushes against the trunk, which is also trying to grow bigger. Eventually the root strangles the trunk which will keep growing wider above and below the girdling root, and may actually encase the root. It also doesn't encourage the tree and shrub roots to expand out in a healthy fashion. Healthy root growth has many agronomic advantages, and it aids in the tree's stability.

So, here are some basics:

Shredded native/Texas native mulches are best when at a depth of 3-4 inches if you are starting from scratch. Even though mulch will break down over time, after two or three years of annual mulching, it often accumulates to unhealthy levels. So start using compost instead of mulch. Or remove a thick layer and begin again. With compost, you get the same weed prevention and moisture preservation, but you get a more accelerated breakdown. So by adding an inch or so per season, you avoid achieving volcano status. If your current mulch is more hardwood shards than natural shredded, consider removing a couple of inches and change to native mulch or compost with your next application.

Re-Mineralizing the Soil

For the past couple of years, on the radio show, many people have heard me promote the use of products like Azomite. It's simply a way to re-mineralize or add trace elements back to the soils. A much-needed addition considering all the rain events in southeast Texas for the past 10 years – Hurricane Ike, the Memorial Day Flood, The Tax Day Flood, Hurricane Harvey & most recently Tropical Storm Imelda. Trace minerals have been washed away with every serious rain or flooding event.

AZOMITE is a natural mineral substance which is mined directly from its Utah desert source. OMRI-Listed for organic production, AZOMITE can be used as an agricultural fertilizer and/or soil amendment product. It is easy, safe to use and good for the environment.

There are other trace mineral packages out there from some great soil amendment companies like Nature's Way Resources and Soil Mender, and they work just as well. Azomite just happens to be the most widely marketed of the trace mineral packages. That means it is the most readily available and also just so happens to come in the largest bags, which sort of makes it the most cost-effective.

If you're still wondering why we need to add trace minerals/ elements back to the soils, here's the best reason of all and a story I've told on the radio show a time or two. The "father of fertilizer," Justus von Liebig developed the "Law of the Minimum" which is important in understanding what AZOMITE does. The Law states that plant growth is determined by the scarcest "limiting" nutrient; if even one of the many required nutrients is deficient, the plant will not grow and produce at its optimum.

So, even if you're building "brand new beds" as this chapter has you thinking, you can and should add trace minerals to the soil, especially if you're not completely convinced you're working with the healthies of rose soils or soil mixes out there.

Chapter 2

Lawn Care a.k.a.

The Ultimate Guide to the

Greenest Yard on the Block

Please indulge me a retrospective look at something I wrote over 16 years ago, when I self-published *Gulf Coast Gardening with Randy Lemmon*. First, I still never get tired of people calling in, posting social media or dropping emails my way that enthusiastically confirm how good THE SCHEDULE works. And for those shout-outs and confirmations, thank you so very much for helping me spread the word that the GardenLine Fertilization Schedule, which you're about to immersed in, is hands down the easiest and most cost-effective way to keep your lawn looking its best.

In 2005, I wanted the intro to the Turf Care chapter of *Gulf Coast Gardening* to pop with some clever writing, hoping to get a wink or nod from the reader who understood the importance of great looking turf in front of the home. I can see a future where the turf in front of lawns may not be as important; however to be blunt, that ain't happenin' in my lifetime.

So, here's what I wrote then, and what I believe is still completely applicable today!

As we step in to the new millennium, it has been said that lawn care is the equivalent to the jousting and duels of yesteryear. You know, weed whackers at 20 paces! Speaking of yesteryear, it has also been suggested that the modern day lawn is the equivalent of a great, green moat. Although you may think of it as 'open' and 'inviting,' our plots of turfgrass essentially separate us from our neighbors. Unlike our European brethren who are very much about fences and hedges between front yards, we are dominated by green, manicured front lawns as a means of separation.

In this new decade of 2020, I want to arm my fellow lawn warriors with the right tools to succeed in the horticultural duel with your neighbors… Dilly Dilly! (some will get that; some will not)

Before we get to the rules and applications of my time-tested fertilization schedule, let me clarify a few things. 1. People call it Randy Lemmon's schedule or GardenLine's schedule, but I've also been very clear that this is an amalgamation of lawn fertilizing advice I've picked up on by several true experts in the fertilizer industry for over 25 years. So, it's not just something I dreamed up out of thin air. 2. The schedule works if you give it one full year. Don't ever tell me you're going to "try and do it this year…" That's an instant excuse out. As the famous philosopher Yoda reminds us: "Do or do not; there is no try!" 3. We make one assumption in knowing this schedule will work – you have to start with some decent soil. (If you don't, no schedule is going to work) If you need to improve that soil before you commit to the schedule, then please read the Soils Chapter in this book again. We cover what to do with really bad soil where there's a need for turf.

Speaking of previous chapters, remember too there's a whole chapter dedicated to Focused Fertilizing, if you need more of the science behind the time-tested fertilizer ratios we recommend for turfgrass in this chapter. Coincidentally, while there have been some minor tweaks to this schedule for the past 20 years you'll still never find the fertilizers, herbicides and fungicides I recommend, and that work perfectly for this schedule at big box stores. You will find them at the nurseries, garden centers, feed stores and hardware stores advertised in the front and back sections of this book, and of course those you hear me endorse on the GardenLine radio show every weekend.

My schedule is mostly focused on fertilizers and pre-emergent herbicides. There's a tiny nod to fungicides and iron supplementation. The schedule, if used as instructed on healthy soils, will do a great job of keeping your lawn green and as weed free as Mother Nature will allow. After you learn as much as you can and commit to this schedule, I'll teach you about all those other ancillary issues: Thatch, runners, insects, post emergent herbicides, insecticides, mowing height and trace mineral supplementation. You will also learn whether my schedule is safe for pets, and why weed-n-feeds aren't safe for anyone or anything.

Stay true to the schedule is all I ever ask. It's technically one visit per month to your lawn. It's really only 11 visits if you do everything. And four of those 11 procedures are considered "optional." Plus there are a few months we can do two applications of different products on the same day, couple that with the two different times you can apply two things on the same day, and I can get you down to as little as six visits per year. I know you can do that! But that's only if you do the "basics" of the schedule, by invoking a couple of those "optional" applications. You'll see when I break this down for you that the "optional" applications are The Early Green Up; Fall Fungicide; Summer Iron Supplementation and Second Trace Minerals.

As an aside, and please don't misconstrue this as a slight towards the Organic Fertilization Schedule, but you must do all those applications for the first few years, if you're truly following the Organic schedule, before you can start 'opting out' on any that are optional. And that means 11 visits or 11 applications without fail.

So without further ado, I give you The GardenLine/KTRH's/ Randy Lemmon's (whatever you want to call it) Lawn Fertilization Schedule 3 Ways! The Original/The Basic/The Organic

THE ORIGINAL

4 Fertilizations (1 Early Green Up; 1 Spring; 1 Summer Fertilizations; 1 Winterizer
3 Pre-Emergent Herbicides (February, May & October)
1 Summer Iron Supplementation
1 Fall Fungicide
2 Trace Mineral Supplementations

THE BASIC

(Opting out on the 'optionals')
3 Fertilzations (1 Spring; 1 Summer; 1 Winterizer)
3 Pre-Emergent Herbicides (Still February, May & October)
1 Trace Mineral Supplmentation

THE ORGANIC

(Be committed to this!)
4/5 Fertilizations (Find the ones that cover the most square footage, please!)
3 Pre-Emergent Herbicides (Yes, there are organic ones; I'm cool with using the Synthetic ones too)
1/2 Fungicide Treatments with Compost Top Dressing (both are optional)
2 Trace Mineral Supplementations (1 for sure is optional)

"THE" SCHEDULE

February to Early March (Optional) – The Early Green Up! Done with fast-acting 15-5-10 formula

February to Early March – 2-in-1 Pre-Emergent Herbicide (Preferably Barricade or Dimension-based ones)

March to April (any time in spring) – Trace Mineral Supplementation. After years of lots of rains and floods, and droughts, we really should be adding trace minerals that help fertilizers do their job. Azomite, Soil Mender, Natures Way Resources and others have bags of trace minerals.

End of March to Early April – Spring Fertilization with Slow/Controlled Release 3-1-2/4-1-2 ratios, such as Nitro Phos Super Turf 19-4-10, Easy Gro Premium 19-5-9 and Fertilome Southwest Greenmaker 18-4-6. Fertilome also has a zero phosphorus formula these days at 18-0-6, but not widely marketed these days.

End of April to May – 2-in-1 Pre-emergent Herbicide (Again, Barricade or Dimension)

June to Early July – Summer Fertilization with same slow/controlled release fertilizers from spring

August (optional) – Iron Supplementation; specific for those using lots of irrigation and seeing a yellowing

September (optional) – Fungicide for Brownpatch

September to October (optional) – Trace Minerals Supplementation (This truly helps the fall feeding)

October – Pre-Emergent Herbicide (Again with Barricade or Dimension)

October (optional) – Fungicide for Brownpatch

October to November – Fall feeding/Winterizer. Use the ratios designed with higher than normal potassium; the third number in the ratios, such as Nitro Phos Fall Special 8-12-16.

Diving Deeper into "The" Schedule

Over the past 20 years, "THE" schedule has been tweaked to add items and specific products. Let me explain. The first 'schedule' ever promoted on GardenLine the radio show was simply two fertilizations (one in spring and one in summer) and the winterizer or fall feeding. My predecessors tried to talk about pre-emergent herbicides, but you needed one for broadleaf weeds and one for grassy weeds, and they never recommended that kind of arsenal three times in the year.

I added the Early Green Up, the fast-acting 15-5-10 roughly 16 years ago, after I was shown the research from 40 years ago that emphasized the importance of a fast-acting 3-1-2 ratio fertilizer coming out of winter and giving lawns a kick in the pants. You didn't have to wait for the April fertilization to make your lawns look good. The one issue that still exists (but you'll know is a non-issue once you finish this chapter, because you'll never be caught doing this no-no) is people seeing 15-5-10 ratio fertilizers that are weed-and-feeds. That's a big no-no for true yard-ners, and that will all be explained a little later in this chapter. There just so happens to be slow-release 15-5-10s out there, but we need the fast-acting ones – in other words the fertilizer only works for about 30-45 days, and gets us to the spring application of the true slow-release ratios that work best here.

As noted earlier, we made a huge change to the original schedule, and it has paid dividends. Again, use a fast-acting, non-weed-and-feed, straight up fertilizer with a ratio of 15-5-10. While there are slow-release 3-1-2 ratios like this on the market, we are looking for instant gratification from a product like Nitro Phos Imperial. That specific formula was designed by turfgrass researchers at Texas A&M University over 40 years ago to help combat the problem of phosphorus lock-up that came with the 13-13-13 basic fertilizer, errantly recommended back in the day.

Say hello to Prodiomine! This was a big tweak to the schedule in the late 1990's. This pre-emergent herbicide joined the industry more than 20 years ago. They found this active ingredient prevented both broadleaf and grassy weeds. We affectionately refer it as Barricade. There were a couple of others marketed about, like Dimension (an active ingredient that is way too long to print here). Then, there was Pendulum, whose active ingredient is Pendimethlin. The problem with Dimension and Pendulum, is that they are extremely hard to find and in some cases ridiculously over-priced. That's why we lean toward the Barricade recommendation 95% of the time.

Our most recent change to the schedule is the addition of trace minerals. Trace minerals and trace element packages have been around for a long time, but in most cases were always recommended for vegetable beds and seldom for landscapes and almost never for turfgrass. The product Azomite hit the scene and sort of changed all that. Azomite and the other trace mineral packages have actually been around for a long time, but it took Hurricane Harvey, several other floods and most recently Tropical Storm Imelda to wake us up to the need for trace minerals to be added back to the soils.

In the past three years, so much rain and so much flood waters have basically leached out so many trace minerals that it may finally help the average homeowner understand what a pioneering fertilizer researcher dubbed the "Law of the Minimum." Again, the law states: Plant growth is determined by the scarcest "limiting" nutrient; if even one of the many required nutrients is deficient the plant will not grow and produce at its optimum. While most of the fertilizers we recommend per The Schedule, focus on the macro-nutrients (the N-P-K), they do offer a hint of trace elements, just not like Azomite and the others do. These trace mineral packages contain rare and abundant trace elements present in volcanic ash in addition to rich minerals present in prehistoric rivers.

While we do say that at least one of the treatments is semi-required per the fertilization schedule, may I recommend at least two applications for the first year you're adding trace elements. You can then cut it back to once per year for the next couple of years, and you can skip a year or two down the line. However, when it rains like it did when Hurricane Harvey unloaded that much rain over a two week period what do you think is happening to the nutrient content of your soils?

Using Azomite or any other trace mineral package with your current fertilization schedule is easy. A 44-pound bag covers up to 12,000 square feet. I would recommend the granular form of Azomite that was developed for broadcast applications. Set your current broadcast spreader to the middle applicator rate. If you apply trace minerals without fertilizer, there is no rush to water the product in. There is virtually no nitrogen, therefore no worries about burning your lawn. If you mix/apply trace mineral packages with your favorite fertilizer, water as you always do. Even if you don't think your lawn ever experienced the flood waters or excessive rains of the past decade, do a re-mineralization of your lawn's soil anyway, because our lawns also lose trace minerals from excessive heat, cold and even a lack of water.

THE ORGANIC SCHEDULE

Before I lay out the 'organic' schedule, let me remind you of a few things you'll notice. First, the dates of application are very similar to the 'synthetic' schedule. Secondly, you will also notice that serious organic gardeners will simply use compost top dressing versus even an organic fertilization, but that's so hard to do 3-4 times a year, every year. So, where there is a need for a fertilization, you'll see "organic compost application" too. Thirdly, we added the trace minerals to this schedule as well. However, and once again, these are completely optional. They may not be needed at all for folks who follow an organic schedule and believe the amount of compost added consistently, is more than enough to make up for trace mineral additions. Lastly, wherever you read about Corn Gluten Meal (CGM) as a pre-emergent herbicide, please know that there is no research at all that proves that CGM really does work on the same level as the synthetic pre-emergent herbicides.

That's why it's worth noting that I have no problem with you using products like Barricade and Dimension in lieu of the CGM. With all that said, here's the organic schedule.

February/March – Organic compost or formulated organic fertilizer. There are so many on the market, but some of my favorites are MicroLife 6-2-4; Nitro Phos Sweet Green 11-0-4; Nature's Creation 6-1-2; Arbor Gate's Blend 4-3-3; Southwest Fertilizer's Earth Essentials 5-1-3; Soil Mender Turf Mate 4-2-3; Medina' Growin' Green 4-2-3.

February – Corn Gluten Meal (as pre-emergent herbicide) ***

March to Early April – Trace Mineral Supplementation (After years of lots of rains and floods and droughts, we really should be adding trace minerals that help make fertilizers do their job) Azomite, Soil Mender, Natures Way Resources and others have bags of trace minerals.

April to Early May – Organic compost or formulated organic fertilizer. (see all the choices listed above)

May to June – Corn Gluten Meal (as pre-emergent herbicide) ***

June to July – Organic compost or formulated organic fertilizer. (see all the choices listed above)

August to September – Specific compost top dressing, as a fungal disease control, specifically against Brownpatch.

September – (Optional) Trace Mineral Supplementation

September to October – (Optional) 2nd specific compost top dressing, as fungal disease control

October – Corn Gluten Meal (as pre-emergent herbicide)

October to November – Organic compost or formulated organic fertilizer as a Winterizer.

*** Remember, in my opinion, you can use the synthetic pre-emergent herbicides, instead of Corn Gluten Meal, because we already know that CGM doesn't work well along the gulf coast. It won't ruin anything 'organically' you're doing with the soil.

And Now for Everything Else

As promised earlier, what follows is as quick write-up on just about every other topic having to do with lawn care for the gulf coast. From mowing to irrigation, to thatch and weed-and-feeds, take a moment and read the "rule" as it applies to you.

<u>Mowing</u>

This has several different categories within: St. Augustine and wide-bladed Zoysias should be mowed with a typical "rotary" lawn mower (think helicopter blade) and should almost always be mowed at the highest level your lawn mower will allow. St. Augustine develops stronger deeper roots and shades out the common Bermudagrass that likes to come in and set up shop on stressed turfgrasses. When mowed too short the blades are exposed to more heat and drought stress. Those are the people that have very yellow St. Augustine grass in the summer months. Bermuda and thin-bladed Zoysias, should be mowed with a "Reel" mower. This is the kind that cuts over the top. Think old-fashioned push mower from years ago, and golf course mowers. If you use a rotary mower on thin blade grass they look sickly and bumpy. St. Augustine should be cut to a height of 3-4 inches, while Bermuda can and should be cut to a minimum height of 1-2 inches.

Mulch-Mowing: As disappointing as this sounds, if you follow the fertilization schedule correctly you will have to mow your lawn. Even so, why should you have to stop to unload grass clippings into bags? That's why I love mulch-mowing. Yes, you can and should bag your grass clippings a couple of times a year, like early spring to help vacuum up the browned grass blades and help prevent a thatch build up. You can and should also use a bagging method when there are lots of leaves from trees on the surface of the turf. Mulch-mowing mostly on St. Augustine grass is extremely beneficial to the yard, because the pulverized grass disappears into the root zone and becomes something of a nitrogen-rich fertilizer. Mulch-mowing can be done on Bermuda and thin-blade Zoysia, but as noted earlier, you really should be using a "Reel" mower, and not a rotary/mulch-mower. In conclusion, mulch-mowed grass is not just good for the typical St. Augustine lawn, it is also the environmentally responsible thing to do, because no more bags of grass to the curb for trash pick-up and eventually landfills.

Aerate/Aeration

Technically speaking, aeration is the naturally occurring process of air exchange between the soil and its surrounding atmosphere. Practically speaking, aeration is the process of mechanically removing small plugs of thatch and soil from the lawn to improve air movement and moisture movement in the soil. Which is why we commonly refer to it as "Core Aeration" along the gulf coast. I'm a firm believer that most lawns can and should be aerated at least once a year. Poorly kept lawns with large square footage of intensely compacted soils (commonly known as hard pan) should probably aerate twice a year for two years, then once a year for two more years, then they can skip several years in between. I ask this question a lot on the radio show: "When was the last time you core aerated?" If the answer is never, then do it twice a year for the next two. If the answer was "a few years ago," get at least one aeration in within the next 6 months. If you answer "every year" then you can start taking a break to every other year.

There are companies that provide this service. Unfortunately the national lawn fertilization companies will make you sign a year-long contract in order to employ their aeration. Yet, there are local lawn service companies that can do this for you as well. Nevertheless, you can rent core aerations machines at big box stores and equipment rental places and DIY.

Compost Top Dressing

Some of you may already be asking, "Why didn't you put the aeration and compost top dressing together?" Mainly because when people are on tight budgets, the aeration has to be done. The compost top dressing is a side benefit (actually a huge side benefit) but one that comes at an extra cost.

I could write a whole book on the benefits of compost as a top dressing in lawns, but I need to keep it short and sweet in this chapter. Over 20 years ago the idea of "compost top dressing a lawn" instead of top dressing with loamy soils or sand, was scoffed at because most composts were chunky and smelled of animal manure. The advent of vegetative composts changed that. Leaf mold composts have now become the standard in the industry and screened in such a way that it looks like dirt. They are the richest looking, microbial-enhanced dirt known to man.

That made it a no-brainer when it came to top dressing a lawn. The soils/mulch/compost producers we talk about on GardenLine all have the best type of composts out there that you can purchase in bulk and/or bags, and do the work yourself. And once again as noted earlier, there are several local companies and landscapers totally capable of doing a compost top dressing. Sadly, there are some national and regional lawn care companies that do aeration and top dressing work, but they notoriously "cut" the compost with loamy soil, in order to save money. Go with local companies, whom we recommend on GardenLine and only use the highest quality composts around.

One other warning about the quality of the compost. There are some soil yards that have compost in bulk, but that are neither screened well nor are they anything close to vegetative compost. If it doesn't look like the richest dirt you've ever seen, avoid it at all possible costs as a top dressing on turf.

Thatch

When I first moved back to Houston to co-host GardenLine in 1996, we did talk a lot about thatch build up. Thankfully, with the advancements in mulch-mowers, aeration procedures, amending the soil with humates and other trace minerals, as well as compost top dressing, I'd have to say that thatch build up is not the same kind of issue it was just 25 years ago. However, there are still problems with thatch, but not on lawns that stay true to my schedule and mow correctly. I don't mean to bury the lead, but maybe some of you are asking "What is thatch?"

Thatch occurs when organic material (in this case, mowed grass blades) is produced quicker than it can decompose. Believe it or not a certain amount of thatch is a good thing. That's because it can work as a cushion of sorts for high traffic areas. Thatch can also be a good insulator from temperature extremes, be that in summer or winter.

For gulf coast homeowners, thatch is that layer of non-decomposed grass blades in the root zone on top of the soil. Too much means moisture and nutrients aren't getting to the root zone as needed. It can be naturally decomposed with amendments like humates, but those who are consistent about core aeration and compost top dressing seldom ever see it. Those who experience too much thatch are those whose lawnmower is not pulverizing the grass to easily decomposable sizes, and those who have likely never, ever, ever core aerated. In those cases, we always advise they rake out what they can, then get the amendment down or get an aeration and compost top dressing done immediately.

Scalping

We simply don't do the true definition of this method anymore! I will leave it at that!

Weed-n-Feeds

The simple answer is never use them! And those who have listened to my radio show and followed us on Facebook or ever followed my lawn fertilization schedule should already know how I feel about Atrazine-based weed-n-feeds. But for those newbies from other states, once again, there is a need to explain in detail. First of all, as the schedule emphasizes, we do our best to prevent weeds by feeding correctly and putting out pre-emergent herbicides. Thus, a healthy stand of turfgrass is the best defense against weeds in the first place. When we do get weed infestations, use post-emergent herbicides that target the weeds and don't poison the soils. That's what Atrazine does. It moves so readily through the soil that it will burn up roots of nearby trees.

I've written several pieces in the past decade asking the ultimate question: Have you ever read the warning labels on the back of a bag of weed-n-feeds? It'll scare you! First, they warn not to use the product several feet inside and outside of the drip line of trees. They also advise wearing a mask, or respirator when applying the product. Yikes! Even if you don't have trees, please understand the chemical Atrazine is very dangerous in our groundwater, posing life-changing reactions in some animals. And lastly, some water treatment plants aren't developed to remove such chemicals. There are Trimec-based weed-n-feeds that I can slightly look the other way if you are bent on using weed-n-feeds, but they aren't always the best all-encompassing weed killer. In fact they only work on broadleaf weeds.

Watering/Irrigation

Let's make this as simple as possible. Healthy grass, because of healthy, organically-enriched soils (see the aeration and compost top dressing chats earlier) do not need more than an inch of water per week on a statistical basis. Mother Nature can provide that on the average, but obviously she doesn't consistently cooperate. So, we have to provide that source of irrigation in the interim. You should do all the tests you possible can to find out how long your specific zones need to run to provide that inch of water per week. The empty tuna can/cat food can test is the simplest. Put an empty can in the middle of the spray pattern of that zone per your irrigation system and run a timer. When the can is filled to the top, look at the timer, and that's how many minutes you need, on average, once a week.

Yes, when things get warmer, we have to do it every 4-5 days, and when temperatures stay in the mid to high 90s and above, we can be watering every 2-3 days. If you're watering every day, no matter what the season, you're wasting water. And even if your lawn seems to be telling you that it needs that much irrigation, then obviously something is wrong with the soil, or you could be using the wrong fertilization program. Higher nitrogen fertilizers, like those found in weed-n-feeds, require lots more water than the slow-controlled release program in this book.

Safe for Pets

At least a half-dozen times each year, I get an email from a new listener to the radio show, or a new follower on Facebook, asking for a "Pet Friendly" fertilizer or flat out asking if my schedule is safe for pets. My response has always been this: First, everyone needs to know upfront I'm an animal lover. In fact, I probably could've, would've, should've been a veterinarian. I went to Texas A&M University with that intention. So, understand that I would never put anything out on my lawns or gardens that I didn't trust to be safe for my pets… or for my kids! Secondly, all the fertilizers I recommend per the synthetic version of my fertilization schedule are made of materials found in common livestock feeds – urea-based nitrogen, potash, and phosphorus. Let me reiterate: They feed that kind of stuff to cows, pigs, goats and poultry. If that doesn't help convince you, or you're still on the fence, either follow the organic schedule, or at the very least just keep your pets off the recently treated lawn for 24 hours. Also worth noting, that some dogs absolutely love, love, love to eat a great many of the organic fertilizers on our organic schedule.

Weed Control

(a.k.a. I'm going to make you an herbicide expert in less than 1 hour)

If you don't want to be that kind of herbicide expert, then by all means skip over the upcoming detailed information. But if you want to know all that I know, keep reading.

When I first started answering garden questions for a living, I felt extremely overwhelmed by weed questions. It seemed impossible for me to be able to identify all these different weeds that everyone was calling in and trying to get an identification on. Then, they needed the right recommendation on how to kill them. It took a few years, but it all started to make sense, and I was able to totally synthesize those kill recommendations once I determined there are really only three weed CATEGORIES, we need to understand. Grassy weeds, Broadleaf weeds and Sedges. Once you know how to identify the weed as one of those three categories, you'll know what to use and when to use it.

There are whole books and websites dedicated to weed identification, and by no means do we have the time or space to cover every conceivable weed to control around here. Here are several varieties of weeds that we endure and how to control them.

Broadleaf Weeds – Clover, Henbit, Chickweed, Dollarweed, Oxalis, Mock Strawberry, Thistle, Spurge, Vetch, Woodsorrel, Asters, Carolina Geranium… just to name a few. If the leaf surface isn't grassy-looking in general, or very thin like a Sedge, then they will all fall into this category of broadleaf weeds. Most broadleaf weed killers will work, but along the gulf coast you have to make sure they are labeled "for southern lawns." Or, when you read the label, make sure your type of turfgrass is listed. It truly is that simple. I'm not really going to try and tell you the active ingredients because they vary and they are all as technical as this: Dimethylamine Salt of 2, 4-D-Dichloropte noxyacetic acid…Geez!!! I can tell you that Fertilome Weed Out and Bonide Weed Beater for Southern Lawns are two of the most highly recommended brands in this market. More importantly, they are the most readily available. But no matter what you use, as long as it has your grass on the label or the label says for "southern" turfgrasses, you still need to add a surfactant to make it stick to the weeds we are spot treating. More on the surfactant in a moment

Grassy Weeds – So, if we know all our grass varieties to be long, slender blades of leaf surface, then it stands to reason that any weed with long, slender blades can be defined as grassy weeds. Crabgrass, Goosegrass, Dallisgrass, Barnyardgrass, Johnsongrass etc. The problem with controlling "grassy" weeds is that the vast majority of grassy weed killers end up killing the turfgrass. There is one organic exception with the powder known as AgraLawn Crabgrass Control. But any other liquid synthetic herbicide labeled for weeds such as crabgrass will kill that which is green that it touches. Still, there are some people that don't mind spot treating the grass weed clusters and when that area is dead, they dig out the dead weed and cover the whole with enriched top soil or compost. This can help heal the actual turfgrass areas that were burnt up with the grassy weed herbicide. When you spot treat with a liquid synthetic herbicide, you still have to add surfactant. With the powder-like controls, you simply use the surfactant on the leaf surface so the powder sticks exactly where you want it to.

Sedge Weeds – All I have to do is say "Nutgrass" and now everyone knows what is Sedge weed. Nutgrass, Nutsedge, Purple Sedge, Yellow Sedge, Kyllinga etc. Grassy weed killers do not work on these thin-bladed weed down to the root, nor will any type of broadleaf weed control. Grassy weed killers can burn the tops down temporarily, but you need sedge-specific herbicides to get control to that "nut" to that root system of the Sedge. The three most readily available ones in this gulf coast market are Bonide Sedge Ender, Gowan's Sedgehammer and Monterrey's Nutgrass Killer II. There are two active ingredients that works across the board. Sedgehammer and Nutgrass Killer II have the active ingredient Halosufuron. Bonide's Sedge Ender has the active ingredient Sulfentrazone. As noted above in both the grass weeds and broadleaf weed sections, it's still critical to add surfactant, even if the product says it already has surfactant – because it's never enough.

Virginia Buttonweed & Doveweed – I've given them their own section in these weed control pages, for several reasons. First, they really are designated as broadleaf weeds, but it's as if pre-emergent herbicides and typical spring and summer post-emergent herbicides won't work on them at all. But, the cool season herbicides and the stronger-than-normal summer post emergent herbicides will work on the newest of growth. Which always, sends us back to the "best way to control" this insidious weeds, is to pull up all you can, whenever you can. Then treat the new growth with the appropriate cool season herbicides in October through December. Or again, pull up what you can in the summer and treat with the myriad of MSA-style herbicides. The problem with most MSM-style of herbicides (Manor, Top Shot, Farenheit, Celcius etc.) is they aren't really designed for St. Augustine yards, so the herbicide contact on grass blades in the area, will cause an awful lot of yellowing. Fear not, because those grasses should come back, as long as you are purposefully spot-treating the weeds, as opposed to spraying the lawn in a wall-to-wall manner. There are other herbicides available, but they are way overpriced for me to be able to recommend them and some require an "applicators license" from the state, the way pest control operators and tree companies do. Once again, the importance of adding Surfactant to these herbicides cannot be overlooked.

The Importance of Surfactants – Here's the detail on why we recommend Surfactants so much. We have hard water along the gulf coast. That's why we need to soften that water, break that surface tension and make the water wetter. I've written for years about the importance of Surfactants, especially in post-emergent liquid herbicides.

Technically, a surfactant is a soluble compound that reduces the surface tension of liquids or reduces interfacial tension between two liquids or a liquid and a solid. It's a linear molecule with a hydrophilic (attracted to water) head and a hydrophobic (repelled by water) end. That's obviously more than most of us need to know. Ultimately, they're important in almost any liquid herbicide. Whether you're killing weeds, unwanted grass or brush, a surfactant is always essential in the mix because most water sources in our area are considered "hard." And hard water tends to bead up and just roll off leaf surfaces.

Try this test: Spray a broadleaf weed killer on some clover, dollar weed or thistle. In most cases you will see that the water beads up. Then, add a surfactant to the mix, and you'll see that it then forms a sheen on the leaf surface. That's the herbicide actually sticking to the leaf and doing its intended job.

There are two ways of adding a surfactant to most herbicides. Just adding a little dish soap to the mix is the simplest way. About a tablespoon per gallon of spray will do. To keep the suds down when using over-the-counter soaps with a trigger-spray bottle or a pump-up sprayer, load the herbicide into the sprayer first, then add the dish soap. Be warned, because adding too little soap won't provide the "surfactancy" needed, and too much soap will create excessive tiny bubbles. That often makes it hard to see exactly where the product is going.

I've discovered, that the new-fangled, super-duper, highly-infused "anti-bacterial soaps" tend to work against the goal of the surfactant. That's why I've always recommended synthetic or professional surfactants like Hi-Yield's Spreader Sticker or Bonide's Turbo. Those two are the most readily available at retail level, and they won't generate suds. I've also used the feed store version of a gallon-sized product called Alligare's Surface. If your herbicide claims that it already contains a surfactant, I suggest adding a bit more. These manufacturers have no clue how hard our water is along the gulf coast. If you only use the ready-to-use bottles of herbicide that hook onto the end of a hose, just fill the void at the top of the bottle with surfactant and swirl. If you use dish soap in a ready-to-spray bottle, you will definitely get lots and lots of bubbles.

Quick Reference Guide for
"The" Lawn Fertilization Schedule.

February 1st – Pre-Emergent Herbicide (Barricade is my choice)

February 15th – 15-5-10 Early Green Up (Nitro Phos Imperial is my choice)

March 1st – Trace Mineral Supplementation (Azomite is my choice)

April 1st – Slow Release Spring Fertilization (Nitro Phos Super Turf 19-4-10 is my choice)

May 1st – Pre Emergent Herbicide (Barricade is my choice)

July 1st – Slow Release Spring Fertilization (Nitro Phos Super Turf 19-4-10 is my choice)

August 15th – Keep an eye out for Brownpatch (I prefer compost top dressing these days)

August 30th – (optional) Iron Supplementation

September 15th – (optional) Keep an eye out for Brownpatch

September 15th – October 15th – (optional) Trace Minerals Supplementation (Azomite is my choice)

October 15th – November 15th – Pre-Emergent Herbicide (Barricade is my choice)

October 15th – November 15th – Fall Feeding/Winterizer (Nitro Phos Fall Special 8-12-16 is my choice)

Chapter 3

Best Trees for the Gulf Coast Landscape

My how the times have changed! When I sat down to write my first book about gardening I had to make a list of the most-often-asked questions which I thought would be a great starting point for said book. That question is still asked a lot around these parts, but it's certainly not the most-often-asked question anymore. One thing that hasn't changed, I still somehow am able to come up with a Baker's Dozen. Which is four more than my original Top 10 list.

The question was: "What's the best shade tree I can plant in this part of the state?" By today's standards, there's so many secondary questions about choices in trees based on size of backyards, proximity to foundations, and simple "aesthetics" to people's properties.

And from that original Top 10 list, I'm still impressed how many are still on it because of those changing standards in the past 20 years. But, I've also had to say bye-bye to a few, as well as introduce a few newer varieties that aren't getting the publicity they deserve. I'll still give you a Top 10 list, but I'll throw in a list of honorable mentions too.

Still, for the most part, I like to answer that question that was once asked so often, based on these three criteria: 1. Fast Growing 2. Will eventually provide ample shade (a bad example would be the River Birch; grows fast, but weak on the shade production) in the shortest time 3. And maybe the most important criteria, it has to be acclimated to our soil conditions. I believe a further explanation is needed on those three criteria: 1. It can't grow too fast, as you'll understand shortly 2. There are fast growing trees, like a River Birch, that would never make a great shade tree 3. If you've read the Soils chapter in this book already, you know what I mean.

And for the record, I've compiled this list with the help of several tree experts in this region. David Williamson from RCW Nurseries and the Williamson Tree Farm. Dave Klineman with US Trees of Texas. And lastly, Martin Spoonemore with Affordable Tree Service. Maybe you won't take my number one suggestion, but if you respect any of these aforementioned tree experts, I will leave you with a quick hit- list of each of these experts number one choice. Later on you'll also learn that pretty much all of us agree right down the line on a few trees you should never consider planting here along the gulf coast.

Shumard Red Oak – (Quercus Shumardii)

Let's be honest, the most popular reason that anyone truly wants a Shummard Red Oak is because of the colors it presents, should we get an early enough frost. It's like you're in New England for the fall leaf change. But as experts have always told me, this is probably the biggest of the fastest growing trees that are acclimated to our soils. And is also said to be the second most demanded tree in Texas behind the Live Oak. All you need to know is that it can get to 60 feet in about 10 years.

Nuttall Oak – (Quercus Nuttallii)

As long as you have ample room between the planting site and the house's foundation, this is by far my favorite tree to meet those three requirements. 20 years ago, it wasn't even close to being my favorite. If it has a downside, it would have to be because of its rather large leaves. That simply makes it a pain in the backside should you have a pool, or if you don't like raking leaves. Otherwise, this close cousin of the Shumard Red Oak does have similar colors in the fall, and definitely its own personality. One of the biggest difference is the size of the acorn. They are larger than most oak nuts and they have a unique striped look to them. So, obviously the fast growth is a huge selling point, and for me the unique acorns are too. For those that can't stand any acorns falling from the tree, this may not be your first choice.

Green Ash -- (Fraxinus Penssylvanica)

Despite the popular claim from 20-30 years ago – ASH IS TRASH – that only ever applied to the Arizona Ash. Sure the Arizona Ash is still well-known for its fast growth, but unfortunately it is not long-lived in this region; usually succumbing a total decline after about 20 years. Well, the Green Ash is considered the next best thing when it comes to that need for fast growth. The few that I planted 2 decades ago have borne that out, and show no signs of cratering. I've told the story before of taking a four foot twig of a Green Ash, and planting it with little soil amendments in the caliche soil of Bryan-College Station; the "Aggie" version of hardpan, clay soil, which I say is five times worse than our clay closer to the coast. In three years, that Green Ash was already 16 feet tall. And since it can mature at 50-60 feet, and adaptable to poor soils, you can see why I've recommended this for a shade tree for nearly a quarter of a century. If there's one down side, it's that it can be one of the last trees to push out new growth early in the spring.

Drummond Red Maple "San Felipe" – (Acer Rubrum)

This tree was barely known 25 years ago, but has definitely become one the most popular additions to gulf coast landscapes, because of its early spring blooms and it's beautiful red colors, if the autumn temperatures cooperate. We've gone from knowing simply about Drummond Red Maples, to Woodlands Red Maples, to the "San Felipe" Drummond, which was found by a local tree aficionado, and propagated in greater numbers in recent years by tree farms like Williamson Tree Farm, affiliated with RCW Nursery. Those experts who grow this "San Felipe" version have grown them side-by-side with other Drummonds and every time the "San Felipe" outperforms the standards.

White Oak – (Quercus Alba)

This is one oak that can get really big. I've seen White Oaks in East Texas that are only 30 years old that are already 90 feet tall. And their span is equally as big, sometimes reaching 65 to 75 feet wide. However, it's not the fastest growing one in the first 5 years. This is actually the one oak I say you have to have patience for if you want that massive shade canopy. Ironically, it's still faster than a standard live oak, yet obviously much taller at maturity. Tell me where you've heard this one before: If the fall temperatures cooperate, the White Oak can give you some uniquely purple leaves during fall months. Side bar: White Oak is still prized for its wood for furniture, and is the state tree of Connecticut and Pennsylvania.

Laurel Oak – (Quercus laurifolia)

This was my all-time favorite tree for about a decade. Yes, it can grow in almost any type of soil, and it grows rather quickly to become that shade tree we covet. Most people mistake it for a Willow Oak, but the Laurel Oak's leaves are a bit longer. Its ability to grow densely, in my opinion, is what makes it so desirable for shade purposes. The only problem with this recommendation is making sure you get a true Laurel Oak. Unfortunately, there are shady characters (see what I did there?) in the tree selling business, and they often, and purposefully, mislabel Willow Oaks, Swamp Oaks and Darlington Oaks as Laurel Oaks. So, please make certain it is labeled Quercus laurifolia on the tag, before you buy anything. Then, you can rest assured you're getting one of the fastest growing shade trees for this region.

Lacebark Elm – (Imus parvifolia)

The Lacebark Elm is native to China and is cultivated throughout the United States in areas with similar hot, dry summers and mild rainy winters. That's us! And that's also why we often call the Lacebark Elm a Chinese Elm. The Lacebark Elm is a medium-sized tree that forms a graceful rounded canopy with long arching branches, which is what makes it a good shade providing tree for our region. The Lacebark Elm produces rich, green and glossy foliage with serrated edges. Of course, it's the exfoliating bark which is what attracts most peoples. Hence the name Lacebark Elm. The bark peels off in a puzzle like pattern and exposes rich shades of gray, green, brown and orange. The two best qualities of the Lacebark Elm would have to be its ability to withstand the harshest growing conditions by growing well in any type of soil; and it's also resistant to Dutch Elm Disease.

Chinquapin Oak – (Quercus muhlenbergii)

I'm not exactly sure why this isn't a top three choice by myself and any of my tree expert friends, because what I'm about to describe to you sounds like it could be number one. Its starts with the knowledge that this is one of the best drought-tolerant, deciduous trees in the state. It grows well in dry, rocky soils in a full sun environment, so it definitely meets that third requirement. It is a medium-to-large growth member of the White Oak family of trees. Normally growing 40'-60' and occasionally as tall as 80', it has an open, round canopy. Although similarity to the White Oaks is obvious, it can take up to 30 years to produce its first crop of acorns, which is why I like it better than the White Oak. The bark also has shallow grooves, and this ash-like look, which peels off as the tree matures making it a striking specimen both in landscape and in the wild. The only reason I can imagine it's not everyone's top choice, is that it's just not as readily available as one would think.

Natchez White Crape Myrtle – Lagerstroemia x 'Natchez'
Bashum Party Pink Crape – Lagerstroemia indica × Lagerstroemia fauriei
Muskogee Lavender Crape Myrtle Lagerstroemia x 'Muskogee'

Some of you may be thinking that Randy's finally lost his mind, that he's recommending Crape Myrtles at all for a shade tree. Hear me out! Let me remind you of our three criteria – Fast Growing; Ample Shade; Acclimated to Our Soils! These are the three fastest growing Crapes out there, and of course if you've listened to my radio show, read any of my other books or taken my advice from social media, you'll already know that we won't be pruning, butchering or murdering these specimens so they can definitely reach a height that will provide ample shade. Now, having said all that, I can prove it to anyone who has ever been and will ever come to my land in Rosehill, Texas! I've got all three of these varieties on my Northern and Northwestern side of my property, untouched for nearly a decade. Others are ready to argue that they'll get covered with insects and then black sooty mold or that they are susceptible to White Powdery Mildew. Not these three! If planted correctly and have a healthy root system, and thus a healthy canopy, they simply don't get insect or disease pressures. These are shade trees, which means, we aren't planting them next to a house or are we planting them near a pool landscape. If we want them as shade trees, they need to stand alone.

Cathedral Live Oak – (Quercus Virginiana 'Cathedral')

Unless you know you're going to be at a certain property for 20 or more years, I've always had a hard time suggesting Live Oaks for people who are looking for that instant gratification of shade. I was able to modify that stance a few years ago, when several Texas-based tree farms starting growing the Cathedral Live Oak. Of course the Live Oak is probably the most popular evergreen tree known for its full, dense canopy, and its consistent height and shape. The Cathedral Live Oak is extremely drought and heat tolerant, and obviously can adapt to nearly any soil type, but it's the uniformity in growth habit that makes it the most desirable Live Oak on the market. I know you've seen typical live oaks as the two trees on a single ¼ acre property, and I know you've seen them on one big property where 20 of them are on each side of a country drive. Admit it: the growth patterns are never the same, whether side-by-side or down a fence line. So, if you just want one Live Oak, choose whatever comes your way. However if you need two for a front yard, or 20 up and down a country driveway, and would like them to grow at the same rate, please choose the Cathedral Live Oak.

HONORABLE MENTIONS

Bradford Pear 'Cleveland Select'
a.k.a. Callery Pear and Aristocratic Pear – (Pyrus calleryana)

I know I've raised some eyebrows and some people's cackles with this suggestion. First, remember this is simply an honorable mention. Secondly, I'm suggesting a very specific version of the Bradford Pear. I'm also suggesting this as a shade tree, not an ornamental. One aspect of this tree's bad reputation, is that it was errantly suggested as a 'small ornamental tree.' So people planted it near foundations and near pools. The flowers aggravated those situations. Some poorer quality Bradford Pears, were produced so fast and in grand numbers, and they were the ones susceptible to Fire Blight, a bacteria disease that made it look like the leaves of a Bradford were hit with a flame thrower. The Cleveland Select and Callery true varieties don't have this problem. So, much like our discussion on the very specific Crape Myrtle varieties earlier in this chapter, if you choose wisely, and you plant it out in the open as a need for a shade tree, Bradford Pears are perfectly fine. In any other situation, I'd have to agree with the critics out there, this is not a wise choice.

Sawtooth Oak – (Quercus acutissima)

The main reason the Sawtooth Oak cannot be on a Top 10 list for gulf coast gardening, is that it can't stand our clay soils, which we need our trees to become established in. However, the more north and east you go, Sawtooth Oaks love the sandier, more slightly acid soils. I have always loved the leaves of Sawtooth Oaks, because of the unique saw-like blades surrounding each leaf. But again, if you live in sandier soils, and you desire to have a tree reach 60 feet in height and 30 feet on its spread, this can be a good choice for you.

Mexican Sycamore – Plantanus Mexicanus

The first time I ever saw at Mexican Sycamore in person, I was spellbound. As of the writing of this book, that same tree that wowed me nearly 20 years ago still exists on the property of TreeSearch Farms in Houston, Texas. I came to know about it, when learning all I could about why the Silver Leaf Maple was such a terrible choice for gulf coast landscapes. Everybody loved the quick growth and the unique silvery color of the backside of Silver Leaf Maples. Obviously, the Mexican Sycamore is not in the maple family, but if you ever wanted a super-tall, super-fast growing tree, acclimated to our soils and with the coolest gray-to-silvery look on the backside of the leaves, this is a great choice. And it's not near as susceptible to Anthracnose as regular Sycamores are. If you want to know why it's not in the Top 10, well much like the Chinquapin Oak, it's not as readily available as all the other suggestions. And since it can get to 90-100 feet in height, I struggle suggesting that kind of tree for a typical suburban landscape for safety reasons.

I WOULD NEVER PURPOSEFULLY PLANT THESE, BUT….

If someone gave you any of these trees for free, and if you follow the tree planting guide (which you can read later in this chapter), then don't look a gift horse in the mouth! But in my professional and personal decision making, I'm not purposefully planting…

Post Oaks / Quercus stellata – because they always have health issues.
Water Oaks / Quercus Nigra – because, they too, always have health issues.
Bur Oaks / Quercus Macrocarpa – because of their huge leaves, and their massive acorns = nuisance.
Corky Winged Elms / Ulmus alata – cool-looking bark, but not so good on the shade.
Hackberry / Celtis sinensis – so trashy, and nasty and so prone to insects.
Sweetgum / Liquidamber styraciflua – because of those spikey balls, which are actually a type of fruit.

TREES YOU SHOULD NEVER, EVER, EVER, PLANT

Just remember, these are trees I would never plant in a typical home landscape along the gulf coast. Then again, I don't think they are good for anywhere in Texas, if I'm being honest. First, never pick a tree for the gulf coast, if it claims it'll grow 10 or more feet per year. Anything that grows that fast, is certainly susceptible to all kinds of problems in the future.

Hybrid Poplars – The only reason I've ever known anyone to purchase a Hybrid Poplar for this region, is because they saw an advertisement on the internet or in a Sunday supplement that totally promotes them on having a tree that grows 10 feet per year. And that may be very true that it grows that quick and tall, but in all cases that just makes it a stick. So where's the shade? While you may be okay with tall and skinny for ornamental purposes, the other reason you won't like these trees for the gulf coast, is that they succumb to so many insects and diseases due to our heat and humidity. They are good for California, just not for any gulf coast region.

Arizona Ash – Humbly put, the Arizona Ash is not a long-lived tree for this region. Sure, I know some folks who brag about their Arizona Ash being over 25 years old and doing just fine. They are the exception to the rule. They did serve a purpose over 40 years ago, when so much of suburban Houston, needed fast growing trees for instant shade. But, when they do start to die, it's a sad and ugly decline. One other reason I detest Arizona Ash was the massive drop of seed clusters in the spring. They didn't compost easily, nor did they mulch mow easily, and that always seemed to leave a grand old mess behind. With the introduction of better Ash varieties, like the Green Ash mentioned early, you still get a rather fast-growing tree, but more acclimated to our soil conditions and with a considerably longer life span.

Silver Leaf Maple – Much like we noted when talking about the Hybrid Poplars, what makes this difficult to write about, is that it truly was a fast-growing shade tree, with a unique color to the underside of the leaves. And that was seemingly the attribute that everyone just had to have. Unfortunately, that fast growth also cursed this tree with a very short life span. Once again, the rapid growth cycle, coupled with our heat and humidity, would often kill these trees in under 10 years. Add to that, when talking about the rapid growth, that the

Silver Leaf Maple's wood became so brittle that it would get damaged in just about any windstorm, and an open invitation to borer insects of all kinds. Thankfully, because of its horrible reputation since the late 1990s, this tree is just not readily available these days. But if you do see it, please don't get sucked in by the marketing or those beautiful leaves with the silver underside.

Italian Cypress – I'm not suggesting you avoid this tree because of its inability to be a shade tree, nor am I suggesting that this won't grow in our soils. I'm adding it to the Never, Ever List, because they are simply not 'designed' for our climate. Unless you consistently supply systemic insecticides to Italian Cypress, (which is environmentally irresponsible) you will eventually lose these tall, columnar evergreens to insects like spider mites. Be honest, haven't you seen more sickly looking Italian Cypress in this region than you have seen healthy stands of this evergreen? Tell me where you've heard this before – these plants simply can't handle our heat and humidity, which is why they do so well in Mediterranean countries as well as more arid states like California.

Carrotwood Tree – You should never have to ever make a decision on this tree. That's because no tree farm/tree purveyor/ tree company worth their reputation should ever sell it here along the gulf coast. The main reason it shouldn't even be considered is that it's considered an invasive plant, and birds drop the seeds everywhere and they sprout up and create a mess everywhere they go. Sadly, there are some "vendors" that I could equate to 'used car salesmen of yesteryear' who will tell you just about anything you need to hear to purchase said Carrotwood. The reason it hasn't had such a "obnoxious spreading" problem in these parts is because they die in the simplest of freezes. Anything below 30 degrees for just a couple of hours, and they are killed. The same kind of shysters, will also try to sell you a Tru Green Laurel, which is essentially a Ficus tree, masquerading as a cold hardy shrub for the landscape. Bottom Line: Avoid both of these, unless you like buying and transplanting new ones every year.

Weeping Willow – For those of us who had one in their own yard as a kid growing up, and thus getting swatted with one for doing something wrong, we'll never want to see another Weeping Willow as long as we live. But for the homeowners, not just the gardeners, in all of us, you don't want it because the root systems are so invasive. They are so aggressive and strong they've been known to ruin underground water lines and crack poured pavement. And much like many of the fast-growing trees we do not recommend, the brittle wood is susceptible to disease and pests.

Number 1 Picks by the Tree Experts:

Dave Klinemen (US Trees of Texas) – **Shummard Red Oak**

David Williamson (RCW Nurseries) -- **Drummond Red Maple "San Felipe"**

Martin Spoonmore (Affordable Tree Service) – **Cathedral Live Oak**

Tree Planting Techniques

For every tree suggested in this book, and even if you want to plant the honorable mentions, the true secret to success is how they are planted. I realize our top 13 suggestions, are highly recommended because they supposedly adapted to our existing soil conditions. But if we give any tree or shrub a fighting chance against our clay/gumbo/caliche soils in southeast Texas, they should all prosper immediately. Simply put, successful root development is the ultimate way to have superiorly healthy trees.

Sadly, most people who aren't familiar with our clay soils often plant the tree in a hole the size of the root ball in question, and slide it in as we say "like a shotgun shell." With our preponderance of clay soils, that is essentially a death sentence for the tree's root system. The young, delicate roots of a new tree, will not be able to penetrate that wall and/or floor of clay. Thus, the root system will either stagnate and give up, or try to grow in just the root ball itself, and often girdle its own root system to death.

I have a technique we have always called "The Twice as Wide and Half-Again as Deep" planting method. If you follow this technique, I promise you are doing everything you can to ensure that the root system of a newly planted tree has plenty of room for the root system to thrive and survive for the first couple of years. By the way, this technique that I'm promoting for trees here, also works on larger shrubs as well.

While most landscape plants in the one-to-three gallon varieties should be planted in raised beds with a rich rose soil, trees need to adapt to the existing soil if you want them to thrive for years to come, and provide that shade they we ultimately need.

We help that root system development by amending the existing clay-heavy soils. We used to be pretty specific and command the use of a permanent soil amendment. Over the years, people have misinterpreted that and thus amend the soils with composts, or mulches and sadly at times they do it with Peat Moss or Sphagnum Peat Moss. To be blunt: forget everything anyone else tells you and add only Expanded Shale as the 'permanent soil amendment.'

The main reason is, because it's the most readily available in our retail marketplace to this day. You're simply blending the permanent soil amendment with the existing soil you've extracted from the ground. Expanded Shale's ability to absorb excess moisture when need be, and its other ability to release it back to the soil when things are too dry, makes it the perfect permanent soil amendment of gulf coast gardener.

The "Twice as Wide" Technique

You dig a hole twice (or even three times) as wide as the root ball in question. And you dig that hole 'half-again' as deep. So, if your tree's container/root ball is 10 inches across, and 12 inches deep – your hole should be at least 20 inches wide and 18 inches deep. All of the existing soil should be placed on a tarp, where you will blend it with the permanent soil amendment, or in our case = Expanded Shale!

Your blend should be equivalent to 5 parts soil to 1 part permanent soil amendment. You could get away with 6-to-1, but that's as far as I would ever recommend. Again, blend all that together and then begin to fill in your 'half-again' portion of the blended soil and soil amendment. You should also tamp down that six inches of the 'half-again portion" so that there's no cavity created, once things settle in.

Once you've solidly tamped down the bottom six inches, then the tree can be centered in the twice-as-wide hole. The same holds true when filling the sides of the "twice-as-wide" portions, when it comes to tamping down the soils as you go.

It's always best to have the tree set flush with the soil line for the rest of the landscape, but if you end up with an inch or so rising above that soil line, you should be okay because there will be some "settling" over time. Then build a temporary berm of soil, so as to make a dam on the outer circle of the dug hole. This allows for the saturating of the root ball each time you go to water.

There is no "one-size-fits-all" watering schedule for a newly planted tree because of Mother Nature's Rains and/or an existing irrigation system. My theories for a newly planted trees is that it will need a good soaking of that damned-up root ball once a week. If rains soak the ground in between, you can skip a week here and there, but never go an entire month without that soaking. The good news in both cases of over-watering and under-watering, is that you've used the Expanded Shale. In both cases the permanent soil amendment corrects the other extreme.

Basic Rules on Watering a New Tree:

1-2 inches of watering soaking the dammed up root ball on a weekly basis for the first 4-6 months

1-2 inches of watering on a monthly basis months 4-12

Once established after a full year, you can leave it to an irrigation system and Mother Nature to provide ample moisture.

Chapter 4

Top 40 Landscape Shrubs

I'm pretty sure I wrote something like this the first time I tried to focus a chapter on Landscape Shrubs… "This topic should probably have its own book." So, in an effort to keep topics streamlined for you, I've focused on the tried and true, time-tested shrubs for this book. That way, you won't be trying plants that are questionable in any way. You will read all these suggestions and know that they are perfect for our climate of extremes. That also means they are NOT prone to fungal diseases or ravaging insect pressures that often wipe out newly planted shrubs or landscape plants.

Since we have a chapter later in this book dedicated to perennials, we tried to keep our focus on what everyone can recognize as a "landscape shrub." Some can be small, some can be medium-sized and some can get really big. They might have a tropical bend to them and there might be a couple of repeaters in the Perennials Chapter as well.

In order to avoid having to apply too many rules and/ or restrictions let's just boil this down to Randy's Top 40 Landscape Shrubs. That means they are my suggestions. These suggestions come from nearly 25 years of doing landscape work, providing on-site landscape consultations, and of course all those years of answering questions on the radio show about what works and what doesn't work along the gulf coast.

Although this chapter will ultimately be about my suggestions, know that most of them are easy-to-care for, and a great many of them are also considered Texas Natives. However, in almost every case, they will need some TLC when they are first planted. Even plants that are listed as "drought tolerant" will still need that TLC the first year. Once they have fully established their root system is when you find out how drought tolerant they can be.

So, no matter what shrubs you choose to plant from the suggestions in this chapter, please understand the need for consistent watering for the first few months, and in some cases the first year. For long-term success with these shrubs it's all about how they were planted, and whether you follow all the rules about raised bed gardening in our region.

I've said it before in the Soils Chapter of this book (and why the Soils Chapter is our very first chapter) and I pretty much say it every weekend on the radio show – "All gardening success begins with the soil. And since our soil isn't that good, how you prepare beds and amend soils is the true key to success."

Texas Wax Myrtle *(Myrica cerifera)* – a.k.a. Southern Wax Myrtle. The Grand Pappy of them all, in my humble opinion. This evergreen shrub is a Texas Native and can be used in full sun to filtered light. It grows fast and stays green year-round. It can be pruned, and should be pruned often to keep it in the perfect height you want. Even when it does get diseases and/or insects they seem to be self-fixing. And, of course, it's very aromatic. They also come in dwarf versions, which are great for smaller landscapes.

Lorapetalum *(Chinese rubrum)* – a.k.a. Chinese fringe flower or fringe flower. Sadly there was a time when many different varieties of Lorapetalum were available, and many of the 'wannabes' couldn't cut the mustard against our heat and humidity. Don't fall for anything with the name Razzleberry or anything like that today. This is the shrub with the distinct, dark marooned-colored leaves that also put on a beautiful display of day-glow pink flowers a couple of times a year. This is the kind of plant that can do well in full sun and/or filtered light situations. What it can't handle is wet feet. You've got to give this plant the proper raised bed drainage it deserves.

Duranta *(Duranta repens)* a.k.a. Golden Dewdrop – Whatever you call the Duranta, I've said before, I don't think it gets enough play in our landscapes. Maybe its popularity hasn't caught on, because it is often mislabeled. Yet, because it is so easy to care for, I still think many people should give it a try. This shrub has an almost relaxed look, with slightly arching branches that will max out at 7-8 feet in height. But it's covered with these blue to dark purple blooms. It's supposed to be tender in colder climates, but none have been lost since the freeze of 1989. Since then, even if damaged up top by any other freeze, they always come back from the root system. By the way, the Golden Dewdrop nickname comes from the fact that it produces yellow berries from time to time.

Cleyera Japonica *(Ternstroemia gymnanthera)* – Simply put, this is the ultimate shrub for shade. As you go through this chapter, you'll read about several plants that thrive in sun and filtered light situations. I can speak out of experience that you won't want to plant this in full sun, however, it is great for filtered light to full shade situations in the landscape. This plant has unique, glossy leaves and a wonderful reddish-bronze hue to the new growth. Most people consistently prune back Cleyera to about 4-5 feet in height, but they can get as tall as 10-12 feet if left alone. But that's not a good idea, because it will get extremely leggy. It's also known as a slow-grower, which means they don't have to be pruned near as much to keep the height in check, and they don't need to be pruned every year either.

Sunshine Ligustrum *(Ligustrum sinense 'Sunshine')* – I was attracted to this plant the first time I saw it at a wholesale nursery in 2012. I asked what it was and they told me Sunshine Ligustrum. My heart sank, because, if this was a cousin of the Wax Leaf Ligustrum (which we never recommend) then it would be susceptible to fungal diseases. Still, I had to experiment with it. The first year, for whatever reason I endured a whitefly infestation, but did what I knew would work and they have survived and thrived ever since. You want them, simply put, for the yellow leaves. This is an ideal hedge in the landscape and offers year-round golden foliage that flourishes in full sun. This sterile, non-invasive cultivar will not re-seed into the landscape. In fact, it doesn't bloom at all, which is good news for allergy sufferers!

Coppertone Loquat *(Eriobotrya Coppertone)* – I've made this statement about Coppertone Loquats for years: If you liked Red Tip Photinias but hate their propensity for fungal diseases, then this is for you. Yes, it can still get fungal leaf spot and some insects like scale, but not to the extent that Red Tips do. At full maturity this plant can get to only 7-8 feet in height, which makes it easy to prune to keep to a 3, 4 or 5 foot shrub. That makes it very easy to prune, mainly because it only has to be done once a year. And of course, the unique color in the leaves is why we truly desire the Coppertone Loquat. You see, it gets that name because of the "copper" colored leaves throughout. It also displays a unique red and orange combination along with the "copper" throughout the year.

Blue Plumbago *(Plumbago auriculata)* – If the Wax Myrtle is my favorite big shrub, and the Sunshine Ligustrum is my favorite small shrub these days, then the Blue Plumbago is my favorite medium shrub. Techinically, it also falls into the "perennial" chapter, but I use it so much for landscape shrubs for so many reasons. The soft, baby-blue flowers are the obvious first choice. But the soft green leaves, which contrasts those blue flowers are equally beneficial as a landscape shrub. The plant is tender to hard freezes, but it always comes back from the root system. In areas where it is protected from winter winds, it always seems to bounce back with some of the first blooms of the season in February through May. If fed the slow-release blooming plant foods that I endorse, they will continue to bloom throughout the summer. Again, while the Plumbago is technically a perennial by horticultural standards, the fact that it can grow to 4-5 feet at maturity is why I use it as a landscape shrub.

Hibiscus *(Hibiscus rosa-sinensis)* -- Okay, again, this is technically not a landscape shrub, but rather a tropical flowering plant. But if we can avoid hard freezes and if we can protect them on freezing nights, you can be rewarded with hibiscus year in and year out that are setting the much-sought-after blooms as early as April. The standard, solid-colored hibiscus (think red, yellow and orange individually) are the ones that can endure light freezes, but I admit the uniquely hybridized ones with multiple colors in one bloom, can rarely be considered a landscape shrub, and have to be thoroughly protected on freezing nights. If not kept in containers and moved for that protection, that is when you lose them. Avoid using "super bloom" fertilizers on hibiscus plants, as talked about in the Focused Fertilization chapter in this book. They bloom on new wood, so give them true hibiscus foods that have higher nitrogen numbers.

Elaeagnus *(Elaeagnus pungens)* – This is one of the more unique and extremely productive shrubs for gulf coast landscapes. But unless you're willing to prune the Elaeagnus consistently, you probably shouldn't incorporate them into your landscape. First, the uniqueness comes from the leaves – silver on the top and bronze-like on the bottom. They grow fast and make a great security hedge, because they can be kind of thorny. The downside is that while they are considered a "fast grower" it comes in a rather sporadic way. It's not the entire shrub that can grow 3 feet every year, instead it's more like a few branches here and there. As noted, because of that unique growth pattern, it truly needs to be pruned several times a year. Research says that Elaeagnus can grow to 8 feet at maturity, but I've seen the unpruned, scraggly growers reach as high as 12 feet on some fence lines.

Japanese Yew *(Podocarpus macrophylla)* – This is another one of those great "all purpose" plants in the landscape. While it is normally recommended for shadier landscapes, I've seen it do just fine in filtered light or lighter shade environments. I've even seen more established ones do just fine in total sun, but newly planted ones don't do well in full sun. If trained correctly, Japanese Yews can and should grow in a columnar fashion. If you need a privacy hedge in a shadier situation, and again if trained properly this is a great alternative. They are coveted for their dark green leaves. Those leaves are thin and elongated, but grow densely.

Oleander *(Nerium Oleander)* – If you have small children and animals that will put anything and everything in their mouths, this is not a good choice for your landscape, because they are very poisonous. But it's so very hard not to recommend Oleanders for gulf coast landscapes. Other than the hurricanes of the past decade, nothing else has taken out the oleanders up and down the main roads in Galveston. They don't need pruning every year, instead once every other year. October is considered the optimum care time for Oleanders. They'll feed on just about anything, and don't seem picky, as long as you're fertilizing it a couple of times .per year. They can be trained to be super drought-tolerant, if you curtail the watering of them starting in July. I am an even bigger fan of the dwarf varieties of Oleanders, however, they are way less freeze tolerant than the larger parent plant.

Nandina *(Compacta & Dwarf mostly)* – I still remember so vividly polling a number of landscapers over 15 years ago to develop this list of "favorite shrubs for Houston." One landscaper wrote back saying "avoid Nandinas at all possible costs." About a year after that submission, he was no longer in business. I've said for years, as long as you stick to the smaller and dwarf versions of Nandina, you've got one of the most uniquely colored landscape shrubs for southeast Texas. The larger versions, which have been known to be a nuisance plant, can spread vigorously, if not constantly pruned or "maintained." Those are the ones that always went by the standard name Heavenly Bamboo. Again, what I love about the dwarf and compact varieties is the unique color combinations in the leaves. You see dark green, red, maroon, yellow orange and copper colors constantly. Then, there usually is a bounty of red berries that will pop in late winter or early spring. And this is one of those plants that actually keeps improving with new hybrids every few years.

Camellia *(Japonica & Sasanqua)* – I included Camellias as a "general" suggestion in the first book I ever wrote, but did not include it in my second book, because I errantly included them in the azalea category of "not worth the effort" plants, if you don't pay total attention to the soil conditions, of iron and acid etc. But in 2011, I bought a property in Rosehill, Texas that was loaded with

Camellias, and still have done nothing to them in over 8 years, and they still bloom like crazy. So, I added them back to this book, but with way more specifics about the two types we work with – Japonicas and Sasanquas. Camellia japonica, commonly called Camellia, and Camellia Sasanqua, commonly called Sasanqua are the genus Camellia but different species. Camellia is a slow upright grower with medium to large flowers that open from mid-to-late winter. That's the kind I have. Sasanqua, depending on the variety, can be either upright or spreading growers with a medium rate of growth and bloom from fall to mid-winter. The flowers on Sasanquas are much smaller than Camellia but the bloom count is much higher. Japonica leaves are distinctively larger than Sasanqua leaves. One other difference in the two is sun/shade requirements. Japonicas grow best in shade to part sun, morning sun being better for them than afternoon sun. Sasanquas will grow from sun to shade but perform best in a sunny spot.

Needlepoint Holly *(Ilex Cornuta)* – Most nurseries and garden centers to this day stock the smaller holly shrub known as Burford Holly. I suggest you settle for nothing less than Needlepoint Holly. They grow faster than Burfords and their pointed leaves are always more supple than the Burfords, and definitely not as dangerous to the skin as the old-style, multi-point hollies. I also love the dark green leaves, so you can plant lighter-colored plants in front for distinct color separation. Needlepoint Hollies work best in shadier and filtered light environments. And they can actually grow to 7-8 feet if left alone. But their biggest attribute for gulf coast gardening, is they can be pruned and shaped much like a Photinia or Ligustrum style hedge. I've also been recommending **Eagleston Holly and Nellie Stevens Holly** if you need bigger and even taller, columnar style hollies.

Pyracantha *(Pyracantha koidzumii)* – a.k.a. Firethorn. In fact, there are so many varieties of Pyracantha on the market these days, I just chose one of the most popular varieties I know is sold in southeast Texas, which is the koidsumii kind. I wish we could go back to calling all of them firethorn. If you don't like the two best-selling attributes of Pyracantha, you'll never like them. They are loaded with berries and they are loaded with thorns. This bold, upright shrub displays intense red berry clusters late in the season and holds them through winter. Because of the thorns, it makes the Pyracantha an excellent choice for use as a hedge, screen, or windbreak with a security bent. No one is running through this hedgerow because of the thorns.

Esperanza *(Tecoma stans)* – a.k.a. Yellow Bells. Tell me where you've heard this one before: Esperanza is technically a perennial, but because they can get up to 5-6 feet in height, I think they make a very unique landscape shrub for southeast Texas soils. I noted 15 years ago, when first writing about Esperanza that you could rarely find this plant 20 to 30 years ago. What makes them so special is obviously the strikingly tubular yellow flowers, which delicately hang like ornaments from this gracefully spreading shrub. If you've ever been to south Texas during the hottest months, this is the one plant that blooms even more the hotter it gets. However, it is susceptible to freezes, which can make them look scraggly during winter months. But even after hard freezes over the past 15 years, I've never seen an Esperanza/Yellow Bells not come back from the root system. They don't like to be over-watered, and you should always remove the expired seed pods, to keep them flowering.

Texas Sage *(Leucophyllum frutescens)* – This is the one shrub, besides the Elaeagnus, that has such distinctly silvery-colored leaves. If a Texas Sage is taken care of properly, it can produce a perfectly shaped, rounded shrub covered with the most striking purple/lavender colored flowers. Once established this is one of the most drought-tolerant plants for Texas. It is often one of the last plants standing, if the landscape had been ignored for years. I will suggest a specific usage for Texas Sage, because while it can tolerate poor soil conditions, and thus be used anywhere, I'm not a fan of it being an entire hedge row. I like it as a stand-alone plant in what you might call an informal and in some cases a Xeriscape-style landscape.

Dwarf Yaupon *(Ilex vomitoria)* – I'm not a fan of regular Yaupon, mainly because it's just not a good looking plant for large shrubbery needed in landscapes. The larger varieties are great for small trees if you can shape them correctly over the years, but seldom do people put that kind of effort into regular sized Yaupons. However, the dwarf versions, while admittedly are over-used in southeast Texas landscapes, they're still a great standard bearer and easily shaped and kept in rounded balls or even as a small hedge row. They simply are the hardiest shrub around. Lastly, you've got to love the Latin name. Vomitoria! So, as a reminder the one I do recommend in landscapes is the dwarf version, which is the one you see "rounded" in so many Texas landscapes. The larger version, which can reach 15 feet at maturity, can also be shaped at the top in a rounded form, but obviously not as easily as the dwarf.

Sweet Olive *(Osmanthus fragrans)* – This is another one of those versatile plants that we need for upright, semi-columnar growing conditions. And of course, it just smells divine when in bloom. Sweet Olives can get as tall as 14 feet at maturity, but it can and probably should be trained to be no wider than 4 feet in width and no taller than 6 feet in height. Most people love to use Sweet Olives as "entrance" plants by the front door and walkways, mainly because of the fragrant blooms. Unfortunately, those awesome smelling blooms only last a few months from February through May. If you want to direct this plant's growth habits in a more upright fashion, they will need to be "tip pruned" all year long.

Natal Plum *(Carissa marcrocarpa)* – Admittidly, I had no respect for this shrub years ago. It looked so frumpy in Galveston, I ignored it as a suggestions for the Houston area. Another reason I paid no attention to it, is because like others mentioned in this book you could rarely find them at local nurseries and garden centers. Ironically, the sandier the soil the better this plant does, which is why it was a standard in Galveston landscapes decades ago. Well, I have a new found respect for the Natal Plum because as I've learned, they truly are so easy to care for, and often won't grow to more than 3 feet in height, and only needs a pruning once every other year. If you're like me, you'll also love the fragrance from the star-jasmine-looking blooms, and the dark green, almost leathery-like leaves. It is well-known to produce a fruit behind that flower as well, and one that's edible. It may not necessarily be the best pick-your-own fruit plant you can have in the landscape when it comes to flavor, but at least it's not poisonous like Oleanders, Lilies and Sago Palms.

Althea/Rose of Sharon *(Hibiscus syriacus)* – Let me be up front, and warn you that Altheas are prone to insect infestations. If you're not willing to treat for and against all manner of thrips, aphids and whiteflies, this probably isn't a plant for you. However, they don't die from such infestations, and as such most people who rarely treat for insects in the summer months, just roll with the punches, no matter how unkept they look. I used to refer to the Althea as a cross between a hibiscus and a crape myrtle. While it technically is in the hibiscus family, the blooms aren't as big as the regular hibiscus and more the size of rose buds. If kept pruned, the Althea is the perfect flowering shrub for the lazy gardener, because it takes full sun and doesn't require much care after it's established – unless you get that insect infestation noted earlier. A fully grown Althea can reach 10 to 12 feet in height. They will produce blooms all summer long and all the way up until the first frost. They will lose their leaves in the winter, much like a crape myrtle, which makes this a better accent plant, rather than any kind of hedge. The variety of colors in the blooms is magnificent, as you can find purple, lavender, white, pink, red, and salmon-colored blooms. Much like hibiscus too, they have single and double blooming varieties.

Knock Out Roses *(Rosa x 'Knock Out')* – The "Knock Out" is simply an improved variety of a shrub rose from years gone by. Some people think they are over-used, meanwhile most people simply love how carefree they seem to be. They are not only easy to grow, they are very disease resistant, which is what totally separates it from its cousins the Hybrid Tea and Floribunda roses, and adds to its appeal. Their bloom cycle is about every five to six weeks. The Knock Out roses are known as "self-cleaning" roses, so there is no real need to deadhead them. However, pruning them a couple of times a year helps keep them from getting extremely leggy. Several Knock Out rose bushes blooming along a fence line or at the edge of an island landscaping is a beautiful sight to behold.

Japanese Blueberry *(Elaeocarpus decipiens)* – I have a 'love/hate' relationship with the Japanese Blueberry. I didn't know much about it 20+ years ago. Then I fell in love with it as a better replacement versus Ligustrum or Red Tips for a large hedge row. I lost that love when they seemed to crater against scale insects, and when freezes in 2010 and 2011 damaged them immensely. So, why am I suggesting them again? Because they just look so good when taken care of properly. Its upright growing and mostly evergreen ability, allows us to use it not just as a hedge row, but as a stand-alone tree or even in a topiary way. Part of its elegant beauty comes from the lush evergreen foliage. Bronze-colored leaves emerge in the spring, which eventually matures to a luxurious shiny dark green color. As the old growth falls, an attractive brilliant reddish-orange color shows itself in the dying leaves. This new growth contrasts beautifully with the old growth. My hope is that if you love them too, you'll keep a watchful eye for scale insects, and if you never want to replace them, you'll protect them on freezes below 20 degrees.

Bottlebrush *(Callistemon citrinus)* – I dare say that there is no other flowering shrub quite like the Bottlebrush. A wonderfully carefree evergreen shrub with masses of showy, scarlet, bottle brush-shaped flowers. A very dramatic accent, hedge or screen plant with leathery gray-green foliage that provides year-round interest. It tolerates poor soils, heat and mild drought when established. That all sounds great, right? Here's the recent problem. They are susceptible to a disease known as Phytophthora. And if you don't catch it in the earliest stages (and we do teach you about that in the Insects & Diseases chapter in this book) you can lose a well-established bottle brush in a matter of weeks. There are dwarf bottlebrush plants as well, and while they aren't as vulnerable to the fungal diseases, they are more susceptible to freezing weather.

Variegated Ginger *(Alpinia zerumbet 'Variegata')* a.k.a. Shell Ginger. This is probably the most dominant tropical shade garden plant out there. Although it was always considered tropical, and by that definition not often considered for year-round landscape purposes, I believe it has overcome that reputation. The striped green and yellow leaves and it's ability to spread, make it like no other ginger out there. I love it mixed with anything else on this list because the contrasts in color are enormous, and because of that I don't just see it as a "tropical" look. Name a shrub mentioned above (with the exception of the Sunshine Ligustrum) and see what I mean. Lorapetalum with Variegated Ginger! Knock Out Roses with Variegated Ginger!! Bottlebrush with Variegated Ginger!!! They work as a border for trees or palms, a filler for a corner bed, or an anchor plant for a mixed garden. The Shell Ginger name comes from it's flowers looking like strands of tiny white seashells. But that's not why we grow them here. Yes, they can look wretched after a hard freeze, but once you cut away all the damaged leaves, they always come back the next spring.

Abelia *(Abelia x grandiflora)* a.k.a Glossy Abelia. I confess that I forget about this shrub all the time. However, this gracefully arching shrub displays bright, glossy foliage backing fragrant, bell-shaped white flowers. It is known as evergreen to semi-evergreen, which technically means you might lose a lot of leaves in the harder winters. You can use it as a low-profile screen or hedge. But I have to tell you that it's the newest variegated Abelia that I think is the stud. They call it 'Kaleidoscope' Abelia. Imagine a typical Abelia but with almost croton-like colors in each individual leaf. I prefer to plant both kinds of Abelia as stand-alone specimens. It provides what I see as small little accent plants here and there when and where you don't have the need to make a hedge row.

Cuban Gold Duranta *(Duranta repens 'Cuban Gold')* -- I knew little about this shrub 20 years ago, then fell in love with it on a trip to Australia, and saw how they use it everywhere to brighten up any landscape. Australians especially used this bright chartreuse-yellow leafed plant as a low-profile hedge rows in between larger more green hedge rows. I fell out of love with it nearly a decade ago, when it seemed like it was prone to severe freeze damage if temperatures got down below 25 degrees – granted, if you didn't cover it. But then fell right back in love with it, when I realized that it always comes back from such freezes, as the root ball never dies. Thus, if you've ever seen my landscapes, they always are peppered throughout, so I can have those bright chartreuse/yellow-leafed balls of landscape shrubs. Cuban Gold Durantas are cousins of the Golden Dewdrop Duranta listed earlier, and while it can get some of those same blue-ish flowers, they are few and far between. You plant this and grow this for the dinstinctly yellow leaves and what they call well-branched. Meaning, it fills out quite well and can be pruned constantly.

Texas Mountain Laurel *(Sophora Secundiflora)* – Who doesn't like the scent of the blooms from the Texas Mountain Laurel? It smells like grape Kool Aid® Let me clarify that this is simply an accent tree for our landscapes in southeast Texas. You don't ever want it to be a hedge row, nor is it ever going to be a big shade tree. Frankly, it does best in more rocky soils of Texas, but if you can keep it from being planted in too wet of a soil, Texas Mountain Laurel grows in limestone soils in Central and Southwest Texas and to 5000 feet in the Chisos and Davis Mountains. This slow-growing evergreen may be grown as a medium to large shrub or trained to a single or multi-trunk tree. The pinnate leaves with their lustrous, leathery upper surface provide year-long beauty, enhanced in mid-spring by the densely-flowered racemes of lavender or violet flowers having the scent of grape Kool Aid®. The black, somewhat constricted seedpods contain red to red-orange seeds which are sometimes used in jewelry. Both seeds and flowers are quite poisonous and contain narcotic properties. In zones colder than Zone 8, flowering is not reliable because of late freezes which damage the buds. Texas mountain laurel is difficult to successfully transplant from the wild. Fortunately it is being produced by a number of growers and is fairly available. Good drainage is a must, as is frequent monitoring for "the worm." The Genista moth larvae, which can decimate the foliage in a few days, is its primary pest. Yellow dye was once made from the sapwood.

Spirea *(Spiraea prunifolia)* – a.k.a. Bridal Wreath. Years ago, you didn't plant azaleas without including Spiraea behind them. That's because they would always seemingly bloom in conjunction with the azalea, or immediately thereafter creating a magnificent color separation, since we mostly grow the white Spirea in southeast Texas. But you won't like the Spirea if you prefer to have a uniform-looking, evenly pruned shrub. Because they look best when allowed to weep over with their branches. They grow in long, leafy strands that produce small bouquets of white flowers. They are stunning. If you prune them, you have to do it once every other year and immediately after the bloom season. If you prune them in the summer or the fall, you will have cut back all that cascading blooming wood.

Muhly Grass *(Muhlenbergia capillaris 'Pink Muhly')* – a.k.a. Pink Muhly and sometimes Red Crystals. The best attribute of most Muhly grasses is that they never get any taller than 18 -24 inches, on the main part of the plant. The flowers can reach to three feet, but it's the main part of the plant we love. Oh, and they're one of the most deer-resistant plants for this region. They are a mounding ornamental grass with fine, dark green foliage; airy plumes of tiny pink flowers. When in bloom in the fall, it gives one a sense of pink fluffy clouds, that will eventually fade to tan seed pods in late fall and winter. I consider them breathtaking when planted in masses.

Aspidistra/Cast Iron Plant *(Aspidistra elatior)* – a.k.a. Cast Iron and Ballroom Plants. Once known as the most indestructible house plant, the Cast Iron Plant has made its way outdoors especially along the gulf coast, as an equally durable landscape plant for shadier environments. I love the name too! You know it has to be tough with the name Cast Iron. A nearly fuss-free, lush, leafy evergreen that will tolerate a range of growing conditions including heat, aridity and dry shade. If I have one complaint about the Cast Iron plant is that they do show browning edges, when they are neglected, because they do require a lot of moisture, and that kind of stress always rears its head during drought-like conditions and when they are competing for the available moisture along with trees in the area. And, yes, it's still an easy care houseplant for low-light situations.

Ligularia *(Ligularia asteralis)* a.k.a. Leopard Plant. Nothing says successful shade gardening like Ligularia. I have a couple of 'friends' in the landscape business, that if the customer has shade, they're going to get Ligularia whether they like it or not. I realize that most people have categorized the Leopard Plant as a tough perennial. All I know is that for as long as I have been gardening I have been planting Ligularia in the shade garden. In general, Ligularia have glossy, deep green foliage with a purple underside. Foliage can be quite large and round to heart shaped. Yet, I still believe that the flowers are what makes them so special. The flowers are orange-yellow to yellow and can look a bit like a black-eyed Susan or along the lines of a yellow delphinium. If there is one down side, it's that they love moisture, and they'll never tolerate a drought-stricken soil.

Aztec Grass *(Liriope muscari 'Aztec')* – I really struggled with keeping this on my list of top choices. But since it really never fell completely out of favor in the last 20 years, I decided it needed to stay. Plus, I needed 40 plants this time, not just 30. The best way to describe Aztec grass is that it's either Variegated Liriope, which I also suggest for landscapes, or it's a less wider and much shorter version of Dainella, which also makes this list. Aztec Grass is an eye-catching, grass-like perennial with green and silver variegated foliage. It makes for a great front border, if you need something other than solid green versions like Liriope or Monkey-grass. It is tough and easy to grow edging for pathways and borders and adds immediate texture and color contrast in mixed beds or containers.

Liriope *(Liriope muscari)* – Here in southeast Texas, we know Liriopi as Liriopi. It is known in northern states as Blue Lily Turf. I'm kind of glad that never caught on down here. So, if you would rather have the solid color border or edging, as opposed to the variegation earlier with the Aztec grass, then you'll love Liriope. I still get a kick to this day when people call my radio show and have

the toughest time pronouncing this plant. I say Low-Rye-Uh-Pee…
but the times that people fumble with it, and say Leer-EE-Ohpp,
gets me giggling every time. This ultimately is so universal and
indestructible, it's no wonder most landscapers use it – maybe a bit
too much! It likes sun, part sun, part shade and shade. If they are
grown in perfect conditions, they will set off a lot of purple, blue
flowers on the tops. But the flowers are not why we grow them down
here.

Dainella *(Dianella tasmanica)* – a.k.a. Variegated Flax Lily.
Remember Aztec grass? If you liked that, you're going to go nuts
for Dainella. I consider this a 'must have' in today's southeast Texas
landscapes, because it simply is a handsome, strappy, green leafed
plant with contrasting yellow or cream-colored stripes that will
brighten the garden year-round. Its tidy clumping habit is ideal for
mass planting near pools, in garden beds and in borders. I do them
in clumps of odd numbers everywhere.

Asparagus Fern/Foxtail Fern *(Asparagus aethiopicus)* I prefer
Foxtail Fern in any situation, but since the two are so genetically
similar I have to lump them together in this book. Having said
that, the Foxtail fern isn't really a fern, but for whatever reason the
nickname has stuck, probably because of the way the stems flow out
similarly to holly ferns. I believe you simply can't kill them. Okay,
I'm sure if you dumped a ton of brush killer herbicide, you could
kill them, but for all intents and purposes this landscape specimen
has made it through freezes, droughts and floods and just keeps on
ticking. This unusual perennial adds textural contrast to beds and
borders. As noted they have sort of a long, upright, plume-like stems
which hold soft, needle-like leaves. They are excellent choices for
hanging baskets and containers. And while they are considered
semi-evergreen in mild winter regions, I've treated mine as if
they are evergreens, because I haven't lost many in all the freezes
we've endured in 20 years. And even if you do lose the tops to cold
weather, they always come back from the roots.

Viburnum *(Walter's viburnum - V. obovatum)* – There are actually a few Viburnums that work in our landscapes, but they aren't anything close to the Viburnums that people from northern states remember. The Walter's Viburnum is sort of the standard in the industry down here and is a semi-evergreen that matures to 10 or more feet, but it can be kept shorter with consistent pruning. The foliage is dense, small and dark green, contrasting nicely with large clusters of small white blooms in spring. The birds are drawn to the purple berries in fall. You'll like the maroon fall color. The Possumhaw (V. nudum) produces olive-green to dark-green, glossy foliage that turns red in fall and drops in winter. This is the Viburnum you want, if you desire the big, shiny leaves. The spring flowers are white; the fruit is a bluish-black and attractive to several species of birds. And this one is a Texas native.

What's Missing? And Were There Truly 40?

If there's a plant you're thinking about and wondering why I don't have it on either of these lists, I promise there's a good reason. Put me to that test: If you'll call the radio show any weekend, I'll hear your reason why you like it and I'll tell you why it was not included. Remember too, that it could be easily prone to fungal disease (see Ligustrum); it could be ravaged by insects (see Junipers and Bulbine); it could be highly sensitive to cold weather and struggle to bounce back (see Mock Orange); or it could be hard to find in this area. Period! I've eliminated Junipers, Indian Hawthorn, Bulbine, Pittisporums variegated and regular, for any of those reasons. Plants I recommend need to be tough enough to handle our heat and our freezes and bounce back with gusto. And although shrubs like Hawthorns and Red Tips can be managed with consistent fungicide treatments, I removed them from this book because that's what I refer to as high maintenance.

AVOID AT ALL COSTS

Wax Leaf Ligustrum
Red Tip Photinia
Golden Euonymous
Barberry
Pampas Grass

I add these two plants to the list I would avoid. However, I will say if someone offers these plants to you free of charge, don't look that proverbial gift horse in the mouth. But I'm not purposefully planting these on my own dime these days.

Indian Hawthorn
Boxleaf Euonymous

I left off two "historical" standards because I also believe them to be two of the highest maintenance shrubs around. Their blooms are short-lived and they are always riddled with insects and chlorosis. Those two are...wait for it!

Azaleas
Gardenias

Chapter 5

Texas Tough Perennials

As I have been known to say over the years, call them low-maintenance, bulletproof, easy-to-care-for, no-brainers, no-fuss ... whatever you call them, you need to add them to your landscape. There are certainly plenty of general lists of perennials for our southern landscapes. But which ones are low-maintenance, bulletproof, easy-to-care-for, no-fuss perennials? Not as many as I would like to see.

What truly sets the suggestions in this list apart from just your average perennial is being able to forget about them once they are planted. Hence the low-maintenance attribute. They also need to have the ability, (I've written in the book *Texas Tough Gardening*) to withstand any kind of insect infestation. As an example, I love Bulbines, but if they are infested with mealy bugs and untreated, they will be wiped out in under a month. Lantana is always a great example of being able to bounce back even if ravaged by spider mites over a summer. Lastly, they have to be able to stand up to our temperature extremes. I don't want a perennial that withers at 100 degree summers, and I need one to put up with freezing temperatures by bouncing back early the next season.

Texas Tough Perennials also need to be available at more than specialty garden shops. So, I give you this list knowing they are available, usually year-round, at the average garden center or nursery throughout the gulf coast. And as we've reminded people in the past, a list like this has to be readily available. Especially for the added color and texture that most of them offer. They don't necessarily have to be in full bloom during the summer. Any perennial can thrive in our spring, but how they survive our robust heat and humidity June through September separates these low-maintenance perennials from the pack. My final criteria: they will flower without mega-doses of bloom-enhancing fertilizers.

I'm sure to get some 'push-back' on this list, or at least questions about why I left off certain perennials. Plus, my list of favorites changes from year-to-year. And we've all figured out over time that there's a whole lot of trial-and-error that goes into building a list like this. Simply put again: This is just my list.

Angelonia a.k.a. summer snapdragon – This still tops my list because it's my absolute favorite. It's one of those true erect little perennials with distinctive toothed-margin leaves and pointed tips, but more importantly flowers that are a vibrant violet to blue. Finally in 2019, you could find more pink and white Angelonias. Those upright stems make the flowers absolutely just stand out on the 8 to 10-inch high stems. The flowers bloom over a long period, from May to the first frost. One more great attribute: if it doesn't freeze too hard, they act like evergreens, bouncing back at the first sign of warmer temperatures.

Yarrow *(Achillea millefolium)* – Selections are available in red, pink, or a white (the white form is weedy) fern-leafed foliage. It blooms spring and fall, is drought tolerant, makes good groundcover, and is excellent for cut or dried flowers. It likes full sun to partial shade and blooms May through July 1.

Mexican Heather *(Cuphea hyssopiafolia)* – It's debatable for some whether this is a perennial or shrub, since it regresses severely after winters. I say it's a perennial. Everyone recognizes it with purple flowers and mounded form. And like most perennials from the cuphea family, they always come back from the roots. It blooms 6-8 months throughout the year and usually mounds at 1 2 feet.

Daylillies *(Hemerocallis)* – OK! I admit this is another stretch in that these are actually bulbs, but for our discussion I consider them perennial bulbs. And they're easy to care for. They bloom in many shades of orange, yellow, pink and red, from April through May mostly and will grow up to 3 feet.

Rain Lily *(Cooperia)* – Again, it's a bulb, but because of its Texas native quality and easy care it fits the perennial category for me. The white and yellow blooms prosper in our summers July through August. It'll grow to 1-2 feet.

Hinckley's Columbine *(Aguilegia hinkleyana)* – It has exotic yellow flowers and delicate foliage. A Texas native, it likes moist organic soil and shade to partial shade. Careful … it's susceptible to spider mites. It blooms March-May and grows to 18 inches.

Louisiana Phlox *(Phlox divaricata)* – With purple to magenta-pink flowers, it blooms in early spring. It's hardy, drought-tolerant and low-growing up to 12 inches. There's a short bloom period February-April, but makes a hardy ground cover year-round.

Lobelia *(Lobelia cardinalis)* – This is often called the Cardinal Flower, but here in southeast Texas it is simply lobelia. In my last book, we also noted that it's a plant that "likes wet feet", so it can be planted in areas that don't drain easily. I see it as a bulletproof perennial above all else. The cardinal-red flowers sit atop unbranched, alternate-leafed stalks that average 2-3 feet high. There are some varieties with pink or white flowers, but it's the red ones that are the major attractors of hummingbirds and butterflies. In this region, I recommend filtered light to partial shade, although in many northern climates it can be a full-sun perennial.

"New Gold" Lantana" *(Lantana camara)* – It has many ever-blooming flower colors, but realistically it's the bright yellow-gold color that most people love. You'll need to shear it occasionally. Heat and drought-tolerant, it loves full sun and blooms spring, summer and fall. It grows to 2 feet.

Katy Ruellia or Mexican Petunia *(Ruellia Brittoniana)* – I'm sure glad they developed this variety from the true Mexican petunia, because I love the dense growth that comes with the ruellia rather than the leggy nature of the original. It has wonderfully purple flowers. It's ever-blooming and invasive, so shear occasionally. It's drought-tolerant, hardy and likes sun. It blooms spring, summer and fall and grows up to 2 feet. The pink and white varieties don't stand up to drought and freeze stresses, like the purple ones.

"Indigo spires" Salvia *(Salvia hybrid)* – This has purple flowers, is very vigorous and likes sun. It can grow up to 3 feet, and you should shear it occasionally. There is also a salvia variety known as "Henry Duelberg" that has come to prominence in the past few years that provides awesome purple and blue spikes of flowers. It is not the variety commonly found at most nurseries, and I should note that their blooms are considered to have more of a blue hue than purple.

Salvia Greggii a.k.a. Autumn Sage *(S. greggii)* – This is another salvia native to Texas, but sort of misnamed since it blooms spring through fall. What sets this truly purple flowering salvia apart from the others, is that it's considered partly evergreen versus the herbaceous perennial nature of the others. This salvia does great in full sun and is extremely drought-tolerant. Greggi varieties are also seen in white, pink and orange beyond the popular and abundant purple versions.

Firebush *(Hamelia patens)* – Its red-orange ever-blooming tubular flowers are prized by hummingbirds. With red fall foliage, it's hardy, drought-tolerant and likes sun to partial shade. It grows to 2 feet, and you'll need to shear occasionally.

"Texas-tough" Verbena *(Verbena hybrid)* – It features pink, purple or red flowers and is ever-blooming. Shear it frequently. It makes excellent groundcover, is drought-tolerant and likes sun. It's susceptible to spider mites and grows about 8 inches.

Turk's Cap *(Malvaviscus)* – Like Mexican heather, it might be a shrub. With unique red blooms and heart-shaped leaves, hummingbirds love them. It blooms prolifically through summer and fall. It can also get really tall. I've see it get up to 5 feet.

Rudbeckia "Goldstrum" a.k.a. Black-eyed Susan *(Rudbeckia fulgida var. sullivanii)* – Needless to say there are actually many variations of Black-eyed Susans out there, but this "Goldstrum" variety is probably known as the most trouble-free. The familiar coneflowers grow upright to about 2-3 feet high, and feature large, daisy-like flowers with the deep yellow rays and dark brownish-black centers. And like most rudbeckia they will handle whatever heat and humidity Texas can deal them.

Cigar Plant *(Cuphea micropetala)* – With orange-yellow, cigar-shaped flowers, it's loved by hummingbirds. It blooms in the fall and loves heat and sun. These are also listed in the drought-tolerant landscape plants section, because they grow more like a small shrub than the typical perennial. Can go to 2 feet.

Dwarf Indian Blanket *(Gaillardia grandiflora)* – It has red-yellow, bi-color flowers and is ever blooming. Remove spent blooms to extend bloom time. It's heat and drought-tolerant, grows up to a foot and loves sun.

Shrimp Plant *(Justicia brandegeana)* – This has extremely unique yellow or red-brown flowers and mostly blooms April through June. Some flower in fall and provide great cut flowers. It likes sun to partial shade and grows to 3 feet.

Trailing Lantana *(Lantana montevidensis)* – Its lilac-colored flowers bloom in spring, summer and fall. It's very drought-tolerant and great as a groundcover, containers or hanging baskets. It prefers sun or partial shade and grows 1 to 2 feet.

Blue Plumbago *(Plumbago auriculata)* – This one features clusters of baby blue flowers. It needs well-drained soil and full to partial sun to reach 2 to 3 feet. You'll also find this one in the landscape shrubs section of drought-tolerant plants, which is how I truly see them. But I didn't want perennial purists to get up in arms.

Coreopsis *(Coreopsis grandiflora)* – Its yellow flowers bloom late spring to summer, and it makes an excellent landscape plant. It reaches 2 feet. (Its Baby, Sun and Early Sunrise varieties are dwarf forms.)

Old Garden Roses *(aka Antique Roses and Heirloom Roses)*- If you want to, you can completely ignore them, which is why they may be the ultimate Texas Tough Perennial. Long before extensive hybridizing, survival without any maintenance was the norm. The down side is that the colors in old garden roses tend to be more muted and pastel. You can't find them at big box stores or mass merchandisers. You will find them at independent nurseries and garden centers, especially those who work with the Earth Kind Program.

Chapter 6

The Importance of Attracting Pollinators

What's the recipe for wooing more butterflies and befriending more bees? To attract native pollinators, an area—big or small—must offer adequate sources of food, water, and shelter. By maintaining natural and cultivated habitats where these insects can nest, rest, and forage, anyone can reap the rewards of healthy pollinator populations.

Some of you may be asking what are pollinators? They are mostly insects when you think about bees, butterflies, beetles and flies (some can be aviary such as birds and bats). They are simply animals that move pollen from male structures (anthers) of flowers to female structures (stigma) of the same plant species. That movement of the pollen results in the fertilization of the flower's eggs. Thus, a properly fertilized flower will produce seeds and the fruit around the seeds, that allows a new generation of plants to be grown. So, that was the more technical way of saying that pollinators are necessary for three-quarters of our major food crops.

Pollination is mutually beneficial to plants and to pollinators. Another benefit is that pollinators receive nectar and/or pollen rewards from the flowers they visit. Sugary nectar provides pollinators with carbohydrates while pollen offers proteins, fats, vitamins and minerals that give them their energy to keep on doing what they do best.

However, in order to attract more pollinators to help with the pollination process, we must plant more pollinator-attracting plants. If we have healthy pollinator-attracting plants, we make for more healthy pollinators and in turn a more healthy ecosystem. Let me also put it one more way for the home gardener – If you want to grow a vegetable garden or healthy fruit trees, you're going to need pollinators.

To ensure that the pollinators will have something to forage from spring through fall, select various natives that flower at different times throughout the growing season. Choosing plants with a variety of flower colors and shapes will also help attract an assortment of insects of varying sizes and habits. Larger clusters of a plant species are better for attracting pollinators than several smaller clusters in and around the garden space. Finally, don't forget the water—all living things need it to survive. Install a water garden, a birdbath, or a catch-basin for rain.

Bees, butterflies, birds, moths, bats, and more! We need all these pollinators to preserve the health of the earth and ourselves. The Million Pollinator Garden Challenge (MPGC) is a nationwide call to action to preserve and create landscapes that help revive the health of pollinators across America. Learn how to participate in this vital effort by choosing plants that attract a variety of pollinators to your home garden by visiting this website http://millionpollinatorgardens.org/

They may have their own list of Butterfly, Bee and Hummingbird favorites, but I'm going to leave you with a very LOCAL list. Keep this on standby when it comes time for you to start stocking your gardens and landscape with these attractors. This first list is a compilation from pollinator friendly lists for gulf coast gardeners. The second, more encompassing list if from a "true gardener" friend of GardenLine, and a frequent guest on the show: Angela Chandler. Angela operates The Garden Academy www. thegardenacademy.com here in southeast Texas, and compiled her all-encompassing list, which will come next.

Nectar & Pollinator Friendly Plants I Can Describe

Abelia x Grandiflora – evergreen shrub, 3' tall x 4' wide, white blooms.

Agastache – "Anise Hyssop", fragrant, many colors, butterfly favorite.

Asters – "New England" and other varieties.

Bee Balm – "Monarda Didyma" and others, many varieties and colors, 12 – 36" tall.

Blackfoot Daisy – 9" tall, white daisy flower.

Butterfly Bush – "Buddleia Davidii", other varieties, dwarf and tall, 18" to 10' tall, butterfly favorite

Coneflower Echinacea – Purple and other varieties, 18-24" tall.

Coreopsis – "Moonbeam" is 12" tall with yellow blooms; threadleaf varieties are 36-48" tall.

Cosmos – "Bright Lights" and others, 24-48" tall.

Daylilies – many varieties and colors, 12-36" tall.

Duranta – All varieties, 36" to 5' tall.

Gregg's Blue Mist Flower – "Eupatorium greggii," Hardy Ageratum, 24-36" tall, blue/purple flowers, butterfly favorite, Texas native.

Gaillardia – "Indian Blanket Flower," many varieties, 12-18" tall.

Gomphrena – "Globe Amaranth," purple/magenta blooms, 12-18" tall.

Heliotrope – blue and white blooms, fragrant, 12-24" tall.

Lantana – All varieties, various sizes, colors and growth habits.

Lemon Verbena – fragrant, 48-72" tall, butterfly favorite.

Liatris – "Gay Feather," "Blazing Star" and others 36-48" tall.

Mexican Flame Vine – 8-10' vine, orange-red blooms.

Penta – all varieties, many colors, 6-36" tall.

Phlox Paniculata – "Summer Phlox" varieties, many colors, 12-

36" tall.

Rudbeckia – "Goldsturm" and other varieties, 12-36" tall.

Salvia Farinacea – "Victoria Blue" and other varieties, 12-18" tall.

Salvia Guaranitica – "Black & Blue," "Amistad", and others, 36-48" tall.

Scabiosa – "Pincushion Flower," blue and pink varieties, 12" tall.

Senna Corymbosa – "Yellow Senna" host for Sulfur butterflies, evergreen, 6-10' tall.

Shrimp Plant – Yellow and red varieties, 18-24" tall.

Texas Star Hibiscus – (and other hardy hibiscus), large crimson or white blooms, 4-6' tall.

Tropical Butterfly Weed – "Ascelpias Curavassica," "Mexican Milkweed," 18-24" tall.

Turks Cap – "Malvaviscus drummondii," 24-48" tall and wide.

Verbena – many perennial varieties and colors, 9-12" tall and wide or spreading, many colors.

Vitex – "Chaste Tree." fragrant blue flowers. Deciduous shrub, 12-15' tall.

Zinnia – all varieties, 6-18" tall, many colors.

Here is Angela's more extensive list, just without the specific descriptions:
Pollinator Plants
Agastache/Hyssop
Alliums
Alyssum
Amaranth
Asian greens (many!)
Aztec Sweet Herb
Bachelor's Buttons
Basil, African Blue
Basil, Common
Bidens
Bishop's Lace
Bluebonnets
Borage
Brassicas – (Kale, cabbage, broccoli, bok choy, mustards)
Calendula
Camellias
Candytuft
Caraway
Catmints
Celery
Chamomile
Chervil
Chives
Cilantro/Coriander
Coneflower
Coral Vine
Coreopsis Cosmos
Crinums (Milk & Wine varieties)
Dianthus Dill

Dwarf Morning Glory
Eupatoriums (White Mist, Blue Mist)
Fennel
Foxglove
Gaillardia
Goldenrod/Solidago Golden Yarrow
Joe Pye Weed
Lemon Balm
Lobelia
Marigold
Milkweeds
Mints
Monarda
Mullein
Nepeta (Catmints)
Oxalis
Parsley
Patrinia
Phlox
Pincushion Flower/Scabiosa
Poppy
Queen Anne's Lace
Radish
Roses (single, open-faced) Rosemary
Salvias
Sedums
Shasta Daisy
Statice
Sunflower
Sweet Marjoram
Sweet Alyssum

Chapter 7

Focused Fertilizers

Even before I started hosting GardenLine in 1996, I too asked the question as a typical homeowner: "Why isn't there one fertilizer for everything?" It sounds rather logical on the surface – an all-purpose, one-size-fits-all fertilizer that I can just throw out maybe a couple of times a year, and everything would be hunky-dory!

Let's get realistic though. If it works for the lawn, it's not going to be good for say roses, right? Let's take that through as many steps as we can. What's good for the lawn, isn't good for roses; what's good for roses isn't good for plumerias; what's good for plumerias isn't good for veggies and herbs; what's good for veggies and herbs isn't good for hibiscus; what's good for hibiscus isn't good for azaleas. You see, I could play this game all day, and while there are some great all-purpose fertilizers I can see using on lawns, trees, shrubs and herbs, that same blend would have the wrong effect on tropical blooming plants like alamanda, hibiscus and bougainvillea.

I've spent the last two decades telling people what to use, what to use it on and exactly when to use it. But I admit in all those years, and four books previously, I've never explained why there are so many different fertilizer blends for so many different plants. Let me rectify that in the *New Decade Gardening*.

Now before the cynic or naysayer inside you says it's all about the money, let me remind you that prior to my hosting the GardeLine radio program, people made their own blends of fertilizers above and beyond the standards that were available. Before mass merchandisers and big box stores with gardening centers, we all pretty much relied on mom-and-pop garden centers and especially feed stores to buy the small array of different fertilizers that barely existed back in the 50s 60s and early 70s.

Back then, if someone wanted a specific fertilizer for a very specific plant, they often had to design it themselves by purchasing the individual elements of nitrogen, phosphorus and potassium. That way, they could add more phosphorus if they needed to feed a flowering plant, as an example. Thankfully, the market saw the need for individual fertilizers for more and more residential customers, as landscapes and gardens took on more significance.

Nowadays, fertilizer sections at nurseries and even feed stores can seem a bit overwhelming to the average homeowner. The shelves are stocked with boxes, bags and bottles with every manner of rose food, vegetable food, hibiscus food, azalea food and yes, even general-purpose fertilizers. And yet, at these well-informed garden centers, feed stores and even some hardware stores you still find all kinds of individual products to blend your own fertilizers to this day – from bone meal, to blood meal, to alfalfa meal to straight potassium.

Regardless of the type, most every fertilizer (organic or synthetic) provides you with the nutritional information about the ratio of nutrients it contains. Prominently featured will be the N-P-K ratio. That's the percentage the product contains by volume of nitrogen (chemical symbol N), phosphorous (P) and potassium (K). So a 13-13-13 fertilizer would contain 13% nitrogen, 13% phosphorus and 13% potassium.

I'm going to oversimplify the feeding of different plants here and hopefully explain why we have and why we need different fertilizers with so many different ratios. Let's start with lawns. The turfgrass along the gulf coast needs more nitrogen than any other element to keep it green. Hence the 19-5-9, 19-4-10 or even the 21-0-0 ratios that are recommended for southern turfgrasses. Interestingly enough, you would think that hibiscus would need a higher middle number to make it bloom, since higher middle numbers like a 15-30-15 would meet that end, right? Wrong! Hibiscus love to bloom on new wood, and the way we get new wood on tropical style blooming plants is to feed it more nitrogen than anything. One of our great sponsors, Nelson Plant Food has a very specific food designed for hibiscus with the ratio 10-4-12. Wow! The middle number, the phosphorus is the lowest percentage. And the reason for more potassium than you would think, is that potassium helps in cell wall protection on all plants. That's also why really good winterizer/fall feeding lawn foods have potassium (the K in the NPK formula) as its highest number. Nitro Phos' Fall Special is an 8-12-16.

For the past 40 years, university researchers have been figuring out what fertilizer blends and ratios work best on everything from azaleas to zoysia grasses (see what I did there? Everything from A to Z). There was a time when 13-13-13 ratio fertilizers were all anyone ever thought we needed. In fact, one of my predecessors Mr. Dewey Compton (I always loved his nickname that my father called him... Dewey Compost) highly recommended 13-13-13 for homeowners on their St. Augustine lawns in Houston.

Texas A&M University turfgrass researchers soon discovered there was too much Phosphorous (the middle number in the NPK formulas) that is was 'locking up' the other nutrients from being used properly by the turf. They discovered that instead of a 1-1-1 ratio that a 3-1-2 ratio worked better at greening up the grass and not allowing for nutrient lock up. That was the introduction of 15-5-10. And to this day, even in my own lawn fertilization schedule, the Nitro Phos Imperial 15-5-10 – that's THE FORMULA introduced by A&M – has been recommended as an early green up by me for over 20 years.

Further research found that instead of having to fertilize every 45 days with the 15-5-10, because that's what its lifespan was in the soil, that they could develop that 3-1-2 ratio with controlled release nitrogen (the first number in the NPK formula) and allow that fertilizer to work for up to 90 days. That was the introduction of fertilizers like Easy Gro Premium 19-5-9 and Nitro Phos Super Turf 19-4-10. This allowed us to cut back on the number of fertilizations significantly. So, as my fertilization schedule calls for, after the early quick-acting 15-5-10, then we only need to fertilize in the spring and summer with the controlled-release 3-1-2 ratio style fertilizers.

Now, there's no way I can break down every type of fertilizer for every type of plant, because I don't have a botany degree. Secondly, I'm trying to keep this book readable and logical. So, let me breakdown a few and hopefully you'll understand why we are "fertilizer focused" here with our *New Decade Gardening*.

Turfgrass/Lawns – All you have to do is follow my schedule (see the Lawn Care chapter for complete details). Remember, nowadays the 3-1-2 / 4-1-2 controlled/slow release formulas have proven best in university research for the past 40 years. Enough said!

Roses – First off, you need to know that roses are known as heavy feeders, meaning they have to be fertilized more often than just about any plant we grow in our landscapes. They tend to do best with well-balanced formulas like 10-20-10 or 5-10-5. The phosphorus is higher, which obviously helps the blooms, but the complete formulas help with new growth and cell wall protection too. Advice on rose growing, as old as time, says to feed every 4-6 weeks usually at the drip line of each plant.

Azaleas – While you can use rose food on azaleas, you truly need to know more about the acidity of your soil. Along the gulf coast most soils need acid added to grow what we affectionately call the Yah-Yah plants (Azalea, Gardenia, Magnolia and Camellia). If there's not enough acidity in the soil, it won't pick up the iron to keep them dark green and growing new leaves. And since they don't bloom year-round, we should focus on healthy green growth. That's why most azalea foods are a very well-balanced like 10-10-10. Even Nitro Phos has a ratio of 8-12-10. So, when we recommend acid-loving plant foods, this is a perfect example thereof.

Hibiscus (and other blooming tropicals) – As noted earlier, while one might think hibiscus fertilizers would be higher in the middle number to promote blooms, it's the first number that's most important. Any blooming plant that gets its flowers on 'new wood' will benefit from hibiscus food. As noted earlier Nelson Plant Food is a 10-4-12. Nitro Phos makes a water-soluble one for the Space City Hibiscus Society that is 18-10-28. And even Fertilome, a more nationally-driven product line, has a hibiscus fertilizer formula of 17-7-10.

Veggies & Herbs – You will often find that most vegetable/herb fertilizers are either completely balanced, such as a 10-10-10 or with a higher phosphorous at the time needed to develop flowers, which eventually would become the "fruit."
A common recommendation for vegetables is to apply one pound of a 10-10-10 fertilizer or two pounds of a 5-10-5 (or 5-10-10) fertilizer per 100 feet of row. When an excess amount of nitrogen is applied to fruit crops such as tomatoes and squash, it is common to have all vines and no fruit. When excessive amounts of nitrogen are applied to root crops such as turnips, carrots and parsnips, you may end up with many leaves and small roots. Hence the need for so many "balanced" fertilizers for veggies. Many all-purpose organic fertilizers, such as Houston's own MicroLife 6-2-4, are great for the crops that need nitrogen enrichment up front to get a lot of green growth that in turn produces the flowers. And since most herbs are all about green growth and no fruiting action, it makes sense for fertilizers such as the aforementioned 6-2-4.

Trees & Shrubs – Most trees and shrubs that we grow along the gulf coast are all about the green growth needed for shade from the trees and filling in hedge rows with shrubs. Hence, we need fertilizers with higher nitrogen numbers. Much like the organics mentioned earlier with the 6-2-4 or even ones like an 8-2-4, that should make sense why the first number, the nitrogen, is the greatest percentage. I prefer organics such as those for just about any tree and shrub, because the organic fertilizers are often packed with trace minerals also needed for the soil and ultimately the tree or shrub's root system. Synthetic tree and shrub fertilizers you usually find with a higher nitrogen ratio, such as Nitro Phos' and their 16-4-8 specific Tree Fertilizer.

Insider Secrets – If the tree is known for flowering action, such as a magnolia, then by all means use an azalea food as we noted earlier. It should also be noted that you can always find insider secrets on other plants to use azalea food on, such as pine trees, and evergreen accent plants such as junipers and cypress. I've also learned that Lorapetalum love azalea fertilizers as well.

Chapter 8

Insects & Diseases

Tell me where you've heard this before. While we could go on and on for this one subject matter, and have my own book on just insects and diseases, we truly had to keep the pests and diseases pared down to the most notorious in both categories. I've been known to say that if it weren't for insects and diseases, gardening would be so simple here. Am I right? All we would have to do is water and occasionally fertilize.

But it's not that easy when you're gardening where heat, humidity, poor soils, winds and rains bring out the worst in insects and diseases. They are a continuous battle. Plus our winters seldom kill off any pest populations, who've found ways to over-winter and wreak havoc the next season. But on the average, it's our temperatures, humidity and extremely poor soil conditions that exacerbate pest populations here.

Speaking of winters, they are usually never severe enough to wipe out naturalized and hardier plants, but ignored fungal diseases can ravage them in a few short weeks. Likewise, our intense summers are really no match for a well-cared for, well-irrigated garden or landscape. However, if anything is ignored during that time frame, they can be overcome by a myriad of insects and proliferating fungal and/or bacteria pathogens.

Bottom Line: In order to control most of the disease and insect pressures, you simply cannot ignore your gardens/landscapes. The old adage about successful gardening "begins with you keeping your shadow in the garden!" And it still rings true to this day. What that obviously means is that an occasional trip around the house, walking and observing, can and should catch problems before they become highly destructive.

Even if you only put your shadow in the garden once a week, that's better than once a month, and even better than ignoring it completely. Yes, even if you have a yard crew that mows and edges once a week, don't use that as an excuse not to walk your garden or landscape consistently.

You see, insects can definitely be controlled easily and quite often very naturally, if you simply take a look once in a while. Early infestations of some of the most problematic insects, like aphids, can be blown to smithereens with a simple blast of a garden hose. If you see a landscape shrub or even a vegetable plant showing spots of any kind, you can usually address a fungal disease outbreak with early fungicides, as opposed to waiting until the whole plant shows spots or lesions many times over.

Insects We Must Control

APHIDS – These insects are often soft-bodied, almost oval shaped, pinhead-sized that huddle together on new growth and on new buds. They also like to gather on the underside of many leaves. They also come in many colors from white, to brown, to pink to green. The most common plants that aphids just love are crape myrtles, roses, hibiscus, camellias and bougainvilleas. They suck the juices from the plant then secrete a honey dew on the leaves and limbs below. That can be the first sign of several insect infestations. And if that honeydew is left alone, that's when you see the black sooty mold that collects on the honeydew. That's when I get the phone calls and emails asking what's the black sooty disease covering my plant? When you see black sooty mold, always know that's a symptom of an insect infestation of some kind. **Control**: Blasts from a water hose during the earliest infestations as well as just about any liquid synthetic pyrethroid insect control. Bifentrhrin, Resmethrin, Cyflothrin etc. Organically, insecticidal soaps are good too, but again, you've got to catch them early for most organic controls to work.

BORERS – I'm mostly talking about the kind that drill holes in trees and larger shrubs. However I will also mention the dreaded squash vine borer too, that decimated cucurbit crops in the vegetable garden. As for trees, to know you have borers beyond the obvious holes in the trunks, you will often notice tiny piles of sawdust at the base of the tree/shrub. A heavily infested tree can die in under two weeks, if the borers are allowed to continually girdle the tree's cambium layer. **Control**: The best known way to control borers in trees is to inject a systemic control, preferably with the active ingredient Acephate. There are many systemic drenching controls with Imidicloprid-based insecticides, but please only apply that control to a tree that does not normally "flower." That's because there has been some correlation with systemic insecticides and the decline in bee populations. The ultimate way to control borers on trees in the long run is to once again take really good care of such trees and follow our deep root feeding/watering protocols in the Tree Chapter of this book. As for controlling squash vine borers in vegetable crops, once you know they've been hit, is to inject a bit of Bascillius Thuringiensis (We simply say B.T. but technically written it is Bt.) into the stem at the bases. If you wait too long, there really is no insecticidal control against such borers. Many people prevent the egg, that becomes the borer worm, from being laid by making a physical barrier at planting time, with a foil wrapped toilet paper or paper towel roll – cut to size so the leaves pop out of the top, but that the moth laying eggs, won't venture past the barrier.

CATERPILLARS/WORMS – We have so many different caterpillars and/or worms that decimate leaves on trees to grass blades in lawns. Some are hairy, some are slimy and all of them are soft-bodied yellow, green or brown in color. Most of them have a healthy appetite for foliage/leaves on a wide ranging choice of plants. And it's a safe bet that if you see any caterpillar on your property, all you have to do is look closely and you'll find a plant (or several) with leaves that have been gnawed upon. From Sod Webworms in the lawn to Tent Caterpillars and Fall Webworms in the trees, they can all be easily controlled and organically at that. **Control**: Once again, you'll see these two words when it comes to any type of worm or caterpillar control – Bascillius Thuringiensis. If you know by the calendar when an infestation normally comes, then be out in front by a week or so and treat the tree, shrub or vegetable plant with liquid Bt. Repeating in 30 days just as an insurance policy. You can also control caterpillars that have to work their way up a tree, by making a ring of Diatomaceous Earth (D.E.) at the base of a tree. Sharp sand can also work as a ring at the base of a tree.

CHINCH BUGS – Remember how we've emphasized the point, that a well-cared for anything is the best defense against insect or disease infestations? This is the ultimate example of that. If you forget to water your turf in the middle of a blazing-hot summer, you will almost always lose your lawn to chinch bugs. These tiny, flea-sized insects are almost always black bugs with tiny white wing pods on each side. They exclusively devour St. Augustine grass. You can rarely see the chinch bug unless you truly know what you're looking for, which is why we suggest performing a water test to flush it to the surface. That can be done by simply laying a garden hose in one spot and allowing the chinch bug, which hates moisture, to run up on the hose. There's also the empty coffee can test, opened at both ends and filled until you see the critters floating to the top. **Control**: Don't let the lawn 'drought out' in the summer months. It truly is that simple. In the meantime, for those who do get an infestation and can catch it early enough, any liquid synthetic pyrethroid (Bifenthrin, Syonara or even Malathion work best) applied three times over a two week period should get control of the egg cycle. I've never been a fan of just granular controls, because they don't break the egg cycle. You can, if you choose, do the granular first, and then the liquid three times still over a two week period.

CUTWORMS – Are usually the larvae of various night-flying moths, but it's the surface cutworm (the larvae) that do all the damage. They get their appropriate name, because they cut off tender plants right at the ground level. They can hit turf, but mostly they hit newly-planted annuals, perennials and vegetable transplants. **Control**: We harken back to the Bt treatment once again, if you know you can be out in front of the larvae laid by the moth. The D.E. control is also suggested if you want to stay organic, and creating that physical barrier with foil-lined toilet paper roles can also work to prevent damage. If you can tell, cutworms need to be controlled at the earliest stages of a "transplant" because you seldom ever see cutworms attack hardier, well-established shrubs, annuals or perennials.

FIRE ANTS – To know the fire ant is to hate the fire ant. Of course, if you've ever been stung/bitten by one, you know they are aptly named. Mostly they are reddish-brown in color and while they don't "ravage" plant-life, like all the other insects noted in this chapter, they can cause a lot of root damage to plants where giant mounds have developed. Another reason we add them to this list, is that they are unsightly too and make a lawn or a landscape simply look bad. They seemingly pop up overnight when the soil is perfectly moist after a good rain. There is no such thing as permanent eradication of Fire Ants, which is why we should always be out in front of them on a control level. **Control**: The Two-Step Method, developed by entomologists at Texas A&M University is still the best overall control. First step is to scatter long-term baits, or long-term granular insect controls over every square inch of turf and landscape. Using products like Amdro or Extenquish or Talstar (which is simply Bifenthrin) over the entire property every 4-6 months, controls those ants wandering around looking for a new home. The second step is to find every possible mound and kill it with a liquid insect control, slowly drenching the entire mound. Some powdered controls like Orthene work to a certain extent in taking down the majority of the mound's population. Some will escape, and that's why you have the first step down to take care of those looking to set up shop somewhere else. I've always had great success following the two-step process, and that became even more important when I moved from a typical quarter-acre sized lot to three acres. It obviously costs more to cover the acreage with the granular first step, but it's worth it. I don't have fire ants pop up relentlessly, as you see on ignored pieces of land throughout the gulf coast. You will also notice long-term control of fire ants, if you can get your neighbors to do the same. It's like they get pushed further and further down the road, if that makes sense?

GRUB WORMS – Normally, if you can keep grub worms from taking residence in the soils of your landscape beds and your lawn, you won't have many June bugs or May beetles hovering around your outdoor lights in the late spring and early summer. But too many grubs in the soil are actually very damaging to the root systems of plants and turf. The best way to describe grub worms is that they are about the ugliest worm you'll ever see. They are curled up, one inch in size, ribbed worms with a hideous looking black head. They are the larval stage of said June bugs or May beetles. **Control**: The problem recently, is that there aren't very many grub worm controls on the market these days. It was easy to control them in the days of Diazinon. There was a semi-natural control up to 2018 known as Halofenozide, which technically was an insect growth regulator, but it too has been removed from the market. The controls you see at big box stores, that show a picture of a grub on the bag are Imidicloprid-based and do work against the youngest of the grub worm larvae.   But research has shown that Imidicloprid is ineffective against fully-grown grubs. The only product still on the market, but that's not easy to find at your average garden center and certainly not at any box store, is known as Dylox, eliminate which has the active ingredient Trichlorfon. Controlling grubs is also a numbers game. One or two, here and there, when working in a landscape is acceptable. Eight to 10 (or more) in a one foot square area, means you have an infestation that needs to be controlled.

LACE BUGS – In Houston, azaleas will eventually get Lace bugs. They are tiny, innocent looking brown bugs with transparent, gauze-like wings (hence the name lace). But they can suck the life out of a plant in under a month. There are two indications to the average homeowner that alert you to a Lace bug infestation. First, the top of the leaves will have a mottled look to them. Secondly, if you look on the underside of the leaves you'll see these caramelized dots, which are the excrement of the bug. Not only do they love azaleas, pyracanthas and mums in the landscape, they're known to ravage several of the trash trees we are well-known for in these parts like Hackberry. And they can do a number on Crape Myrtles, Italian Cypress, Cherry Laurels and Water Oaks. **Control**: Most of the liquid insecticides we have already mentioned will work on knocking down the populations. To keep them from doing much damage, you should consider a systemic control, especially on the azaleas, with liquid Acephate products. Systemics with Imdicloprid can work well too. Organic gardeners swear by Nicotine Sulfate as the ultimate Lace bug control, but that too is hard to find.

LEAF MINERS – I really shouldn't be adding this to this book, because it really isn't life-threatening but rather unsightly at best. And while there are scores of leaf miners in horticulture, it's the citrus leaf miners I talk about and how to control them. You can't tell if you have leaf miners by looking for an actual insect, as they are microscopic and embed themselves in the leaves of citrus plants, leaving behind a ton of squiggly lines and quite often curl up that leaf. While they aren't life-threatening, nor would I term them as "ravaging" like many of the aforementioned insects, untreated leaf miners can reduce yields over a period of time. **Control**: To this day, the same treatment I have promoted for nearly 20 years is the best method. It's a two-step approach that we apply during the spring growing season, immediately after the early spring pruning. Once the new growth starts to emerge you treat with Neem Oil one week and the Spinosad the next week. Do this alternation for a total of 8 weeks (longer if you so choose). The best part about this control is that both products are 100% organic. Hopefully, this protocol has shown signs in helping prevent citrus psylliad-type miners, which are the culprit in spreading the dreaded Citrus Greening disease.

LEAF ROLLERS – Personally, whenever I get leaf rollers on my Canna Lilies, I simply shuck them out and prune back the damaged leaf. But since this is also from the caterpillar family of insects, as we explained earlier, the Bt (Bascillisus Thuringeinsis) is a great way to control these critters that damage mostly lilies and many other tropical gingers by doing exactly what their name says – they

roll up the leaves and live inside there feeding on the leaf tissue.
Control: So now you know why I control them by shucking out the
caterpillar. Just spraying a Bt will do nothing because of the rolled
up leaf. However, if you want to prevent them from rolling up a leaf
of a lily or a ginger, treat in advance. Also, dusting with a powdered
form of Bt (commonly found as Dipel) after the shucking is a great
long-term control. I choose not to use any systemic controls on
lilies or gingers because of the pollinator populations like bees and
butterflies that really like the pollen and nectar of blooms.

MEALYBUGS – Oddly enough, mealybugs are from the scale
family of insects. But unlike scale, these oval-shaped insects
covered with a white, cottony wax will move. They tend to cluster
in groups and when that happens it is often mistaken as a white hair-
like fungal disease of some kind. However, when they do cluster
up, they can multiply rapidly and can deteriorate a plant quickly.
Control: It's much like our discussion with aphids earlier, meaning
the simplest way to control these is with a blast of a garden hose
early each morning. You have to be careful with insecticidal use,
because products like Malathion will damage the plant. Bifenthrin is
probably the safest insecticide, when you can't use Malathion.

SCALE – Since scale insects don't fly (the parent insect can sort
of hop around) and moves very little, few people every really know
they have a scale insect infestation. But in my opinion it is the most
insidious insect we have to endure, because it can set up shop on just
about anything. Here, we mostly see them as tiny, white specs, with
a hard, waxy outer shell that sits up and down the middle spine of
most leaves and congregate on the stems below. Some scale, such as
tea scale, infests the underside of many different plants, and again,
are seldom detected until it's often too late. There are some scale
that are brown, and I've even seen some pink-looking scale. But
they all have that same hard outer shell and when they all do their
job sucking the life from the leaf, they secrete that honey dew, which

eventually becomes saturated with the black sooty mold again. The best description I've ever heard, from a novice gardener was that it looked like someone had spattered tiny dots of paint all over their holly bushes. **Control**: To this day, the best control on scale insects, whether it's on shrubs or trees is to spray with Malathion. First, know the difference between live and dead scale, because where there are dead scale, you have no reason to treat with a chemical. In fact, the dead scale, white or brown, will slough off in due time. To determine live scale versus dead, simply pluck off a few of the scale dots and squeeze one with your finger tips or finger nails. A live one will ooze out a liquid that comes in many colors. Dead scale, white or brown, will be crunchy with no visible liquid of any kind. Another way to determine life in a scale insect, is to take a whole leaf covered with the white specs and rub on a piece of white paper. If they are alive, you'll see streaks of yellow in most cases. A dead scale will show little to anything while swiping down the white paper. Almost every recognized organic insect control does little to kill off scale, probably because of the hard outer shell. However, using a true dormant oil spray in the months of December through January is a viable option.

SLUGS AND SNAILS – These critters have been known to devour entire leaves overnight, but as least we don't have as serious a problem with them as they do in states like California. For me, a slug is just a snail without a shell. And they both leave slimy trails. To anyone that has never dealt with either, the slug is about a one inch long, with a slick, slimy, black to greenish-gray colored bodies. They both love to feed at night, again devouring leaves and/ or leaving Swiss cheese-like holes everywhere. **Control**: Once you detect any kind of slug or snail damage, the control methods always work, if you'll be proactive and jump into action. Most slug and snail baits are pet friendly and can put up with rains and irrigation, meaning that a one-time application is usually the fix. I prefer a two-step approach, with the help of good old-fashioned Diatomaceous

Earth (D.E.). I sprinkle bait everywhere there's soil or mulch, and then make a ring of D.E. at the base of the plant, because they have to crawl from somewhere they were hiding during the day. The D.E. tears them apart, like you and I crawling over shards of glass. The bait gets to those that have never made it to the stem of the plant. A fun, almost 7th grade science-experiment-way of controlling slugs and snails is with a pie tin full of beer, laid level with the soil or mulch, so they can walk straight in. They are attracted to the yeast, and within a minute or so, they bloat up and drown. Just throw all the drowned ones away the next morning. DO NOT use excessive amounts of salt to kill the slugs and snails, because while it does kill the pest, the concentration of salt is bad for the soil and the root systems of nearby plants.

SOD WEBWORMS – For a few years, this was the bane of most homeowners with St. Augustine lawns in the late summer and early fall of 2013 to 2018. These worms were the larvae of the moth that hovers just a few feet over the lawns. Much like trying to control chinch bugs, these worms if not treated properly would eat up a lawn in a matter of a week. If you didn't detect the tiny white moths by walking through a lawn, the next thing to look for are tiny half-inch sized worms mostly green but with touches of brown throughout. If you detected neither the moth of the worm, then you would also notice this sort of "wave of death" creeping from one side of the lawn to the other. **Control**: Putting out granular insecticides rarely does the job. And again, like the chinch bug control noted earlier, it takes several applications of liquid insecticides over a two week period to break the egg-laying cycle. However, since it is a worm doing all the damage, this is where the tried-and-true organic control with liquid Bt really comes in handy. Remember Bt targets only worms and caterpillars.

SOUTHERN PINE BARK BEETLES – These are tiny, black beetles that are no bigger than a grain of rice. They make tiny intrusions on the outer bark of pine trees. A first indication can be tiny piles of saw dust at the base of a pine tree, but once those holes have been bored, then quite often the hole will ooze a tiny bit of sap after the original sawdust removal. The pitch, or ooze, coming from these intrusion holes is usually a yellowish to reddish-white resin mass. Pine bark beetles usually only attack pine trees that are under some kind of duress. Another way to sadly determine pine bark beetles damage if you've never paid attention to the trunks, is that the pine needles will turn yellow first, before turning completely rust-colored and falling. **Control**: It hurts to note, that once a pine is heavily infested with pine bark beetles, it's almost impossible to salvage. If you are fortunate enough to detect it in it's earliest stages, any number of insecticides from Malathion to Bifenthrin to Termiticides can be sprayed in and around the tiny holes. I once again recommend a systemic insecticide with Acephate for "injection into the bark" for long-term control. And if you choose to be extraordinarily proactive and want to prevent pine bark beetles from doing any damage, then any manner of systemic tree formulas also work like Imidicloprid-based formulas.

SPIDER MITES – I do think it's important to note that spider mites aren't actually spiders, but relatives of spiders. To the naked eye, if you can get a good close-up look of these miniscule creatures they look like little dots of red, yellow or black. But that's just it! These itsy-bitsy creatures are undetectable to that naked eye in most cases. You often figure out you have them when you start seeing chucks of evergreens just turning brown and seemingly overnight. They tend to love the evergreens like Junipers, Cedars, Cypress and Arborvitea. And they are notorious for ravaging flowering plants like Roses, Lantana and Azaleas. And although tiny in size, they are massive in number when slowly and surely damaging those evergreens, one browned-out section at a time. One of my favorite things to do when showing people that they have spider mites, is to tap a section of the plant down to something like a piece of white paper (or flat piece of cardboard) and see the those critters plop down on the paper and start scattering everywhere. And when you can see them on the white paper, you will notice they do look like tee-tiny little spiders. Sometimes early in the morning, a tiny web may be seen thanks to early morning dew. That too can be an early indication of spider mite damage ahead. **Control**: There are some natural 'miticides' on the market in the form of Spinosad-based insect controls, but often that's just a contact kill for that one spot you see. While the product is still available to us, I once again recommend Acephate-based systemic insecticides, mainly because unlike the Imidicloprid-based ones, Acephate can also be a contact kill along with the systemic property it is well-known for.

STINKBUGS – 20 years ago, I didn't consider the stink bug enough of a threat to put on this kind of list, but the more and more people wanting to grow their vegetables (especially tomatoes) the more and more we see this annoying pest. These miniature, tank-looking pests can pop their beak into so many tomatoes that they can ruin a crop in under a week. **Control**: They have to be removed. There is not a spray in the world that works effectively against Stinkbugs. That's why removal is the key to success. I always say this on the air when talking about stink bugs, and it's often mistaken as a joke. Control: Get out in the tomato crop with a portable shop vac, and vacuum them off the plant.

THRIPS – Much like when we talk about spider mites, thrips are fairly imperceptible to the naked eye. But these tiny little boogers can absolutely mangle soft, budding flowers and new leaf tissue, mostly on Roses and any things in that family of plants, such as Hibiscus, Altheas and Rose of Sharon. They can also damage flowering plants on the flower and from the inside burrowing further into the bud. And they are difficult to eradicate because we don't want to be using systemic insecticides on flowering plants, so as not to mess with the precious pollinator populations of bees, hummingbirds and butterflies. **Control**: When it's a bad infestation, you will have to opt for liquid insecticides every couple of months. If my plant is devoid of flowers, I can use a systemic in the root system and Bifenthrin as a contact kill on the tops of each plant.

WHITEFLIES – I still think scale insect is the most insidious one on this list, because people tend to ignore it. I used to think whiteflies were the worst. These life-sucking insects fit their name so perfectly. They are tiny little version of flies that have white bodies and white to translucent wings. They will fluff up in the air when you make contact with them on a branch or a large leaf. They'll settle right back down for continuation of their dinner plants. Like aphids and scale, whiteflies also secrete a honey dew which in turn becomes that black sooty mold. They love vegetable gardens to crape myrtles to several evergreen shrubs like Viburnum and Sunshine Ligustrum. **Control:** In the past, and why I thought they were the worst insect we could endure, is because there wasn't a single insecticide that could get the job done. What we've learned in the last 20 years, is that you have to alternate between two to three different insecticides. They simply don't become used to a single kind of insecticide. Of course, the systemic controls we have discussed throughout the whole chapter can and do work, but you don't want to use systemic insect controls on vegetables and any other flowering plant because of the threat to our pollinator pals. But where you can use contact insecticides, start with one of these and every 3-5 days, hit them with the next one and so on, until you know there are no more whiteflies fluffing up on that plant again. I suggest any two or three of these, just make sure the label doesn't warn against certain plants. Bifenthrin, Resmethrin, Cyflothurin, Syonara and Malathion.

Pillbugs, Sowbugs, Earthworms, Leaf Cutter Bees, Leaf Cutter Ants Earwigs, Silverfish, Centipedes and Millipedes can cause a bit of damage here and there, especially if completely ignored. But I don't personally consider any of these highlighted insects that much of a threat to our love for gardening. So, they don't get that kind of special attention in this book.

Diseases We Need to Control

I may not get as much agreement on this theory from the average homeowner, but I think we are blessed to not have too many fungal or bacterial diseases that are blatantly destructive in this growing region. Sure we have our fair share, as you'll read in a moment, but fungal diseases are way easier to determine that hidden insect infestations, so much so that we can and should be treating for and/ or against early signs of fungal diseases. There truly is only one fungal disease in trees that, even if detected early, is untreatable and that's Canker. Everything else you're about to read can be detected early and treated immediately, with huge success rates in curing such diseases.

ANTHRACNOSE – Even if you know you have Anthracnose in your trees, and even if you don't treat right away, seldom is Anthracnose a life-threatening fungal disease. When a tree starts looking sickly and the leaves of said tree show irregular brown blotches and cause premature leaf drop, that's when you might have the disease. There are often situations where the leaves have lots of browning edges too. Often when I get to see those leaves, I can usually tell whether it's just a water deficiency or Anthracnose. When an expert gets their eyes on such leaves, they will look closely at the middle vein of a leaf and look for a brown dot, somewhere up and down that vein, as if it's clogging or blocking the nutrients from getting to those outer edges. **Control**: Good old-fashioned Chlorothalonil (you may know as Daconil) has always been a standard in the tree industry, but today, we also know that copper-based fungicides as well as systemic fungicides with Propiconizol (we often refer to as PPZ or Banner-based) fungicides can do the trick too. If you keep the trees on a consistent deep root watering/ feeding schedule as noted in the chapter about trees, you can often keep Anthracnose at bay.

BLACK SPOT – This is a fungal leaf spot type of disease mostly found in roses and that particular family of plants. It does what it says, by causing these black to brownish spots that eventually turn a leaf brown to yellow and then fall off. If black spot is not controlled early, it can defoliate a plant, but again is not life-threatening right away. However, a rose with a scant amount of leaves prevents the plant from manufacturing its food effectively and there usually is a slow death involved. Black Spot is definitely a product of our humidity. Where you have humidity and you're growing roses, you're going to have black spot. **Control**: Experienced Rosarians know to spray their roses on a weekly basis usually April through November. In the colder months with the least amount of humidity there is no need to be out there on a weekly basis. Less than two decades ago, there was an introduction of fungicides with systemic properties that have longer lasting residuals, meaning they can be used every two to three weeks. I also know of rose-aficionados that spray their roses early in the mornings with a product known as Consan 20 (used to be Consan Triple Action 20) simply using this proprietary blend of alcohols, to "wash off" the dew that might bring about the black spot. Along those lines, if you want to be as organic as possible, is to provide ample amount of aeration. That means around the bush, and ample space between each rows and each plant, and keeping them pruned in an open vase form, then the humidity won't be collected in a certain area. It also helps to continuously remove infected limbs and fallen leaves from the area.

FUNGAL LEAF SPOT – This is what affects all those plants I no longer want in the landscape – Red Tips, Ligustrums, Hawthorns. It starts as small brown to black spots that ruin an entire leaf yellow. The spots can get bigger and defoliate a plant over a period of time. It's technically not life threatening, unless you do nothing for several growing seasons, which simply makes them unsightly. **Control**: Now, having said all that, if you're willing to be very proactive with any of the aforementioned systemic and or copper-based fungicides above, then you can easily control this disease.

GRAY LEAF SPOT – This is very specific to the lawn, and very specific to St. Augustine grass. Most people, unfortunately, don't notice it until it has caused a huge area of yellow grass. If you can catch it early enough, you'll simply notice what seems like fungal leaf spot circles in the middle of grass blades. Technically, they become diamond-shaped spots. Lesions begin as tiny, round or oval gray to brown or black spots on leaves. Spots may be surrounded by a yellow halo or general chlorosis with purple to brown borders. If you put enough of those "spots" together you get the yellow hue over a larger area. So, people that don't diagnose it correctly end up putting out fertilizer iron and exacerbating the problem. Gray leaf spot may be showing up because of nighttime watering, frequent rainfall, high humidity, heavy dew (i.e. prolonged leaf wetness), plus rapid, lush growth courtesy of recent fertilizations. Lawns with severe gray leaf spot have areas that seem to just fade or melt away. **Control**: The best fungicide to this day doesn't have a label for Gray Leaf Spot, but we know it works. It's Daconil, or the Chlorothalonil-based fungicides at 4 ounces per gallon of water in a pump up sprayer. Leaves may be blighted gray, usually from the tip downward. Besides the commonly found Daconil, the fungicides Banner, Banner-Maxx, and Heritage are also approved for use on gray leaf spot, although they're harder to find and often more expensive.

BROWNPATCH – You know when someone from another state gets brownpatch for the first time because they always describe them as "crop circles." The more you use compost top dressing from now on, the less you will have Brownpatch. But for those that aren't there yet, Brownpatch forms circular patches of brown to yellow discoloration that grows larger and larger if left untreated. This problem usually happens when there's an unholy alliance of moisture, cool nighttime temperatures and fertilizers blend together. They are always prevalent in areas that also have drainage issues. The nighttime lows between 60-68 degrees seem to the biggest culprit, along with any night time moisture. You may notice that areas that don't have proper drainage are the first areas to show the circles. If there's any good news is that Brownpatch is not life-threatening. Once you do have it though, it is what it is, and you can't green it up until the next spring. **Control**: I'm not a fan these days of hitting Brownpatch with a ton of chemical fungicides. That's because once you start the regimen, it has to be done every 21 days until the weather is cold enough. That's why I say, get on a program where you are doing compost top dressing every six months for the next two years, and I promise you won't see Brownpatch down the line. Still, for those that want to spray or apply fungicide (and liquid forms are always best in my book) Banner-based, Propiconizol-based, Myclobutanil, and Terrachlor/PCNB types of fungicides all have a label for Brownpatch. If you can catch Brownpatch spores with a preventative dose of one of these fungicides before they blow up in size, I'm all for that. But once you see it, it's worth reminding you to be treating every 21 days until our low temperatures stay in the 50s and below from November until February.

TAKE ALL PATCH – Everything I just wrote above can be applied to Take All Patch as well. Yes, I will give you the details in the meantime. I remember writing 15 years ago about how whether TAP was a new phenomenon or were we all just misdiagnosing other issues over 20 years ago. Still, the sad truth is most people don't even know they have TAP until it's too far gone and the grass has simply melted away. When the disease is active, the first symptom is a yellowing of the leaves, almost like you have an iron deficiency. Then there's the gradual thinning out of green grass. But if we can all go out and do a simple test once we see any sort of yellowing grass, what happens next will always tell us whether we are dealing with TAP or that myriad of other things that cause the grass to yellow. The Test: Grab a handful of grass and pull it up. If all the roots and dirt come with the grass you almost assuredly have TAP. If the roots stay in place, and all you get are grass blades in the hand, then we would then try to figure out whether it's any of the other maladies. It's also still confusing to this date via research articles just exactly how a yard gets TAP. I equate it to a cancer for the yard, because if detected early and treated properly, you can stop it in its tracks. If it's ignored it will kill the entire yard, and if you don't replace all the soil or treat that soil correctly for future turf, it's systemic in the soil and will wipe out new turf as well. **Control**: High-end compost is the answer. Especially the Leaf Mold Compost or Vegetative Composts as noted in the Soils Chapter. You can try any and all of those aforementioned fungicides as well, but nothing fixes this cancer of the soil better than top dressing with compost. Again, two times a year for the next two years is essential. And a once-a-year compost top dressing will keep it from ever rearing its ugly head again. The good fungi and the good bacteria in the compost eat up the bad bacteria and bad fungi in the TAP. So, think about this too before you ever run out and apply gobs of synthetic fungicides to control TAP. They eliminate all the fungal spores and all the bacteria, even the good ones that keep a soil naturally healthy.

FIRE BLIGHT – Like most fungal diseases that have all kinds of technical names, but that we give perfect description names to (see Brownpatch, Rust, Take All Patch) I've always thought Fire Blight was the most appropriately named disease, because it gives a swatch of tree leaves a burnt-looking appearance. It's almost as if someone came by with a flame thrower to the outer leaves of the tree. Fire Blight is technically a bacteria, that can be hard to control because it can be transmitted and transferred by birds, insects and pruning equipment. Fruit trees and non-fruit bearing ornamentals, such a Bradford Pears are still the most highly susceptible plants to Fire Blight. **Control**: And considering this is bacterial in nature, there simply aren't a lot of 'bacteria-cides" out there. There are still Streptomycin-based products on the market, which are few and far between and not on the cheap side either. We also used to recommend Consan a lot, but they changed the recipe. I'm not sure that the elimination of one of the three ingredients that gave it the "triple action" is the one that was considered a bacteria-cide. I've never been able to get a clear answer from the present day producer of Consan 20. But I've also never heard anyone say it didn't work on Fire Blight, since the change. I still feel confident in that recommendation.

POWDERY MILDEW – Coincidentally, the product I just mentioned is still the best way to control Powdery Mildew on anything other than food crops. Yes, good old Consan 20 still makes a great fungicide and it simply washes the mildew away, and hopefully leaves a day or two of residual behind to keep it from popping back up. For those not used to our humidity, you need to understand what Powdery Mildew is. For those who have Crape Myrtles or grow Roses, there's a strong chance you already know what it is. Again, the name is perfectly descriptive, because it is a white to gray-ish, powder-looking mold that coats newly formed leaves on plants. Powdery Mildew stunts the growth of the plants and can, if left untreated, can kill some plants in the absolute worst case scenario. **Control**: Quite often, Powdery Mildew can be prevented as long as air circulation is improved. There are 'mildew-resistant' varieties of Roses and Crapes, but if the perfect concert of poor air circulation, new growth and just enough of the fungal spore in the area, even those kind of plants will get the disease. So again, the Consan 20 is the easiest thing, because it simply washes the mildew away, and is probably the less environmentally caustic of all the fungicides mentioned throughout this chapter. I actually know of rose growers who go out every morning there is dew on the plants and rinse them down with Consan 20 just to prevent anything and everything. But I also know Crape Myrtle owners who, when they see even a hint of Powdery Mildew, they spray these flowering trees with any of the systemic fungicides noted earlier. I also know that the organic method using Neem Oil on a weekly basis, can also keep Powdery Mildew at bay.

PHYTOPTHORA – This has been the hardest fungal disease to control in the past decade. I admit that even I have misdiagnosed this fungal disease as Cotton Root Rot (CRR), but I know the difference now in how Phytopthora loves to keep eating away at new growth, while CRR seems to burn out sections of mature growth. By the way I've found the disease spelled phytophthora, but we'll stick with phytopthora. So, flowering shrubs and small trees are especially susceptible to this disease, and again all their new growth. In southeast Texas, Bottlebrush (Callistemon) and Japanese Blueberry (Elaeocarpus) in recent years have been hit hard. Basically, phytopthora is a root-rot disease (again, similar to cotton root rot) brought on by the wicked combination of a wetter-than-normal fall followed by a wetter-than-normal late winter. **Control**: I also believe that any systemic fungicide with the Phosphorous Acid as its active ingredient is best known for phytopthora control. Names like Garden Phos, Agri Fos and Reliant, are well respected in the horticultural community. Here's another big difference in the two diseases, and how they are controlled in my personal opinion. You can use the Phosphorous Acid fungicides as a part of the soil drench. However, if you want to do the two-step approach (you'll read about for CRR later), then by all means feel free. Another difference between phytopthora and CRR, is that you have to treat for Phytopthora for two to three months. Cotton Root Rot usually only needs one treatment. In fact, it should be on a monthly basis until new growth is no longer ravaged by the disease. One final thought from an expert in the industry: Make sure you remove mulch volcanoes around any of these plants if you want to control pythopthora, and spading fork the area so it breathes better, and quite literally getting to the root of the problem — where the phytopthora actually exists. Lastly, if you plan on replacing any plants that died from this root rot, be sure to drench the area fully before replanting.

COTTON ROOT ROT – Tell me where you've heard this story before. Much like the Phtophthora as diagnosed above, CRR give plants the look that they are dying a section at a time. However, CRR is very plant-specific too. And CRR, since it is truly emanating from the roots has to be treated both in the soil and on top of the plant surface. The smaller evergreens like Yaupons, Hawthorns and the bigger ones like Red Tips, Ligustrums and Wax Myrtles are the ones that show CRR over phytopthora. And as noted above, CRR seems to eat up mature sections at a time versus the ravaging of new growth from the phytopthora. CRR comes after we've had what I've always referred to as a "roller coaster ride" of moisture. One month really wet, and the next month really dry. If we put a few months of severe ups and downs together, we see CRR. **Control**: My recommendation has always been a soil fungicide, specifically Captan for the roots of the affected plants and the nearby ones, and then a systemic fungicide like Banner-based or Propiconizol-based fungicides for all the leaves on all of the plants in the area. This disease moves through the soil very easily, so the soil-based fungicide is critical.

RUST – I actually love to answer questions about the fungal disease Rust, because I always get to play the old school analogy game. If it looks like a duck, walks like a duck and quacks like a duck, then it's probably a duck! The same can be said for rust. Everyone knows what rust looks like on metal objects, right?! So, if it looks like rust, and can rub off the plant in between your fingers and gives that rusty, orange color, then it's probably Rust. Rust affects mostly flowering plants in this region. Roses, Plumeria, Lilies, Iris and Azaleas. **Control**: Almost any fungicide out there, and any fungicide I've noted throughout this chapter has a label for rust. I'm fond of Consan 20, but Mancozeb and any Copper-based fungicide are considered standards in the industry. However, cultural practices are actually the best way to prevent ever getting Rust. If you see it early enough, remove all the infected parts and destroy them. If you can avoid having water splashing up on leaves this helps, as well as removing excessive amounts of mulch in the area. Lastly, if you want to treat organically, so to speak, then consider dusting your plants early in the seasons with agricultural sulfur.

Chapter 9

10 Best and 10 Worst Things

to Happen to Gulf Coast Gardening

(Since I took over as GardenLine Host)

A garden-writer friend asked me and other "supposed experts" in the industry to come up with our Top 10 list of the BIGGEST CHANGES -- horticulturally speaking -- we have seen in the past two decades. When she first asked, my mind couldn't even contemplate more than five. However, I asked around and even asked my radio audience, and so many more came to the surface. At that point, with all those comments, it occurred to me "changes" could mean good or bad. Here we are a couple of years later and I figured there were so many GOOD changes and still so many BAD changes, that I would make two separate lists for this book. The Top 10 Best and Top 10 Worst things to happen in our Gardening/Horticultural world. I give it to you in good old-fashioned, David Letterman style moving from 10 to number 1.

Top 10 Best Things to Happen in
Our Gardening/Horticultural World

10. **My Fertilization Schedule Works.** There have been some minor tweaks over the past 20 years, but nothing works better along the gulf coast to keep lawns looking green and weed free. It's hard to believe that national lawn fertilization companies don't even try to replicate it, but thankfully a couple of local companies do. All I ask is that you start by having decent soil beneath. If your soil stinks, no schedule is going to work, period! And the only other request is to stick to it for one full year. All I'm asking you to do, on the average, is one thing per month. Really it's only 8-9 visits over that 12 months.

9. **Deep Root Feeding/Watering of Trees.** The more drought-like months we endure, the more obvious the need for deep root feeding/watering of trees. For years, several supposed

tree doctors and "certified" arborists claim that such deep root anything is useless to the tree. That's because they are trying to sell some kind of inject-in-the-cambium-layer of the tree type of service, which just so happens to be three to five times more expensive than typical deep root feeding from honorable tree companies. A tree expert I trust, reminds us all that you're feeding the root zone, revitalizing the root system so that it will recycle correctly and provide the nutrients and minerals the tree needs to have a healthier root system and thus a healthier canopy.

8. **We are Planting More and More Fruit Trees.** This is especially true for citrus trees along the gulf coast. It helps that we have learned that a typical backyard can handle a dozen fruit trees, and that space is not a limiting factor these days. That's because most fruit trees can and should be pruned to a specific size and shape each year. I have a friend in a typical 1/4 acre size lot in suburban west Houston that has over 50 different fruit trees on his property. The average person that listens to my radio show, has anywhere from 4-6 different fruit trees. What has also truly helped is that some nurseries sell fruit trees year round. Just 20 years ago, they were only for sale in the early part of the year and that was that.

7. **More and More Texas Native Plants are Used in Residential Landscaping.** Thirty years ago, when people started promoting the need for more and more Texas native plants as part of the landscape, there wasn't much of supply in local nurseries and garden centers. The advent and growth of mass merchandisers and big box stores with nurseries didn't help that cause at all. Thankfully, that much-needed marketing push has gone way beyond people in the country needing deer-

resistant plants these days. So many specialty nurseries and garden centers throughout the state promote their forte towards Texas native plants. And besides the deer-resistant benefit, people are learning about the low maintenance and drought tolerant benefits as well.

6. We are Getting Better at Attracting Pollinators. Some of that credit goes to stories in the news about bee colony collapse and decline in those populations, as well as issues with less butterflies and even less birds. Nowadays, there's so much education and information sharing about the need to plant the right plants to attract more pollinators, which ultimately benefits our ecosystem. More and more people are paying attention to this issue. In fact, I dedicated an entire chapter in this book to plants that attract Pollinators.

5. Organic Insect & Disease Controls Have Improved. We've been introduced to a whole new world of Organic/ Natural Insect Controls – say hello to Neem Oil, Spinosad, Plant Oils and the like. I used to poke fun at a local gardening show 20 years ago that recommended a Cedar Oil product for every insect issue out there. Nevertheless, Cedar Oil is a great deterrent, but not an organic contact kill. But we've gone way beyond only cedar-based products in this modern day of organic pest control. And there will be new ones introduced on a faster clip than any new synthetic insecticide in future years.

4. The Internet. If everybody I talked to on GardenLine was "internet capable" this probably would be my #1. Let's face it though, for what I do, the internet has been extraordinarily helpful in getting answers quicker to people, and now via social media (see: Facebook **https://www.facebook.com/ GardenLineWithRandyLemmon/**) the sharing of valuable information is phenomenal. We were still filling "Self-Addressed-Stamped-Envelopes (SASE)," requests just 15 years ago, but not anymore. For the consumer in gardening, not only can they find answers quickly too, they can order everything from seeds to tools online these days.

3. **Organic Fertilizers Have Gotten Better.** They aren't just smelly-chicken-poop-clouds-of-nastiness anymore. They are more advanced in that they don't smell so awful, can be used in broadcast spreaders and are more cost-effective. They still don't work as quickly as synthetic fertilizers, but in this day and age of environmental awareness, it's great to see organic fertilizers evolving with the wants, needs and desires of the marketplace.

2. **All Kinds of Soils.** Thirty years ago, if you wanted rose soil (one of the best, "getting started" soils for all landscape and outdoor potted plant purposes) you had to make it yourself. Equal parts loamy soil, sharp sand and compost/humus material. All kinds of quality soils by the bulk or the bag are so readily available these days, there's no need to make your own. There are very specific soils from Azalea Soil to Citrus Soil to Blueberry Soil.

Drum roll please….. The #1 Biggest and Best Change for me in our gardening sphere here along the gulf coast.

Compost/Compost/Compost. Better Compost for so many uses, period! Compost as top dressing; Compost as mulch; Compost to build veggie gardens… you name it. Plus, more companies are making higher quality composts by the bag and by the bulk. Remember 30 years ago, bags of compost were clumpy and smelly. Today, compost is better and more readily available.

Top 10 Worst Things to Happen in
Our Gardening/Horticultural World

10. **The Annual Crape Myrtle Massacre.** At this point in my "GardenLine Life" it seems like a battle we may never win. Too many landscape companies use the butchering back of Crape Myrtles to keep crews busy in the winter months. Still, there's so many homeowners that prune back to the nubs every year because they were ill-informed years ago, or they see uneducated landscape companies doing it. If you've always pruned them back each year, give it a rest for this next year. See the difference in the health of the Crapes and you'll learn it's not worth the massacre every year. If it's a size issue, then you've got the wrong size Crape in said spot. And, yes, it's still called Crape Murder as well!

9. **Zoysia Should Be Our Turfgrass of the Future.** I can't believe Zoysia has not taken over as our predominant choice entering the year 2020. It was hailed as the "grass of the future" over 25 years ago. And I thoroughly agreed with that. A southern turfgrass that would need less fertilizer, less water and no garden chemicals for fungal diseases or insect pressures. It's still available and I do know more people are installing it, but why hasn't it taken over, so to speak? Probably the cost. Even though you're eventually saving the money with less water, fertilizer, fungicide and insecticide uses, the fact that it's nearly double the price per pallet of grass is what's kept it more on the sidelines.

8. **New Homebuilder Landscapes are Simply the Worst.** Let me repeat: New homebuilder beds are simply the worst! I actually have an uncle in the landscaping industry, who does

quality work for one particular homebuilder, unfortunately he's the exception to the rule. Most homebuilders, for whatever reason, think that this is the best place to save money and go with the cheapest bids out there. That usually means scraping up the dirt/clay on which the home was built, and building their version of raised beds, and plopping in the cheapest plants they can get their hands on and still upcharge like it's a quality plant. Then, they plug them into that horrible excuse for a raised bed and cover it up with mulch; and even then, use dyed mulch which is further poisoning the soil. I suggest the builder hand you the landscape allowance and you add to it whatever it takes to come up with 1% of the home's value. A $240,000 home should have at least $2,400 invested in the landscape. Build beds with only high quality soils and high quality mulches, and what you choose to plant is your decision, not the landscaper looking for the cheapest plant, so his upcharge makes him even more money.

7. **Weed Killers Have Not Evolved.** Weed Killers are basically the same they were nearly 25 years ago. While I may be stretching the "worst" definition on this example, it certainly is the most disappointing trend, if you will. I still can't believe that some university researcher has not come up with an all-purpose weed killer that's safe for most of our southern turfgrass. Sure we have non-selective herbicides that kill everything, and selective/targeted herbicides for broadleaf and sedge type weeds. I'm also baffled to this day, that there is no such thing as an "organic" all-purpose broadleaf weed killer. While something like AgraLawn Crabgrass can take out grassy weeds and a scant few broadleaf weeds, there is nothing on the organic market that is an all-encompassing broadleaf weed killer, like Bonide Weed Beater for Southern Lawns.

6. **Mowing Practices Have Not Evolved.** This is especially true for St. Augustine grasses, where people mow too short on a consistent basis, and wonder why their turf looks so yellow and sickly. How many years have we pleaded for St. Augustine lawns to be mowed at the highest level a mower can go?! As for Bermuda and thin-bladed Zoysias, why are you growing those if you're not willing to invest in a Reel Mower? That's the kind of mower that cuts over the top, like they use on golf courses.

5. **Mulch Volcanoes.** Every time I see mulch volcanoes, I say a big fat UGH. Ignorant and/or uneducated landscapers do it, especially on new builder homes, and then the homeowner just continues down that same sad path. We only need a couple of inches of mulch around the base of newly planted trees, not a foot or more in height. This is a death sentence to trees the older they get, as the roots try growing in the mulch pile versus the ground below. By the way, trees should eventually show root flares the older they get. Mulch should never be a permanent thing for trees. And stop trying to plant flowers in the mulch ring as well.

4. **Compacted and Tainted Soils.** This has been an issue since all the heavy rains and floods from Tropical Storm Allison to Hurricane Harvey to the Tax Day Flood to the Memorial Day Flood. When waters cover up lawns and landscapes for days at a time, nothing good is happening to those soils. We talk in-depth about Soil Remediation after flood waters in the "Building the Perfect Beds" chapter in this book, if you're unfamiliar with Soil Remediation Protocols. While most of the "Worst Things" in this list are sort of man-made or self-induced, this is a problem caused by Mother Nature, and one that's hard to prevent. But when it does happen, we know how to fix it.

3. **Weed-n-Feeds with Atrazine.** I keep shaking my head in disbelief and disgust at this one. And I've been doing that for nearly 20 years. There has been plenty of opportunity in the past two decades for this product to be removed from the market by EPA, FDA, USDA and any other alphabet soup government agency with such power. I don't have time to get into the "why it hasn't been removed" in this chapter, but I'll remind you, like we did in the Turfgrass Care chapter, why it's so bad. It kills trees and it contaminates ground water. No matter how good you think Atrazine-based weed-n-feeds are at killing of weeds and greening up of the grass, it's the negative things you don't see that make it such a heinous product to use on any residential lawn.

2. **People Don't Do Their Own Lawn Care Anymore.** I'm famous for saying the phrase: "If you neither have the time or inclination you hire things out." Nevertheless, because people don't do the simple work themselves – mowing, trimming, mulching etc. – folks end up hiring the nearest and/or most convenient landscape company working in their neighborhood. I would estimate that only one out of every 50 landscapers has any kind of education/knowledge on proper care practices. This is why we see so many diseases and weeds being shared from yard to yard. This is why Crape Murder happens with regularity. And this is why dyed mulch is so unnecessarily prevalent. And this is also why weed-n-feeds are improperly used as well. If you want things looking good and staying healthy in your landscape consistently, and if you still 'neither have the time or inclination' to do the work yourself, at the very least vet these landscapers out. Meanwhile, take back control of your lawn/landscape by doing the work yourself.

Drum roll please…... The #1 Worst Thing to Happen to us and our landscapes over the past 20 years…

1. **Dyed Mulch.** There's certainly more detail on why Dyed Mulch is so, so bad for our landscapes in the Soils Chapter of this book, but let me remind you of a couple right here. Mulch should first and foremost provide three beneficial elements. 1. Reduce Weeds 2. Conserve Moisture 3. Add Organic Matter Back to the Soil. I also think it should look natural. Black and/or Red Dyed mulch doesn't look natural at all, and none of them are helping build back organic matter to the soil. They are almost always made of chipped up wood from pallets and discarded timbers, and then dyed. Tell me how that's helping the soil at all? And even if it is dyed with something "organic" it's still a dye, and it's still leaching into the soil. If you don't do the mulching yourself, as noted in #2 above, then ask your landscaper to start using a more natural or native hardwood mulch. If they refuse, then they should be fired immediately.

Chapter 10

Month to Month Checklist

When I wrote my first book, I told readers this was "the best" part of the book, because it was easy-access information. This is now my fifth gardening book and the fourth self-published book. And I admit as my writings have gotten more mature and stronger, while this is not "the best" part anymore, it certainly is the chapter with the most reminders.

While there are new entries in each of the 12 months covered, there there are a lot of repeats from the first book I have to admit. But that's still a good thing, because I want you to be referring to this book at least once a month. It can be at the beginning of the month or at the end of a month, looking forward and thoroughly looking through the next month's ideas. At the very least it'll jog your memory about some things you should be thinking about doing for the coming month, and maybe it'll also remind you to re-read other chapters hopefully for a second or third time.

It could be about lawn care, tree care, when to plant certain things per the calendar or when to be on the lookout for certain diseases or bugs. Bottom line: I want you to be referring to these month-by-month to-do lists, so much so that you've dog-eared several of these pages.

I don't want these lists to frighten anyone. In fact, if you read through these lists, and feel overwhelmed, just take heart and remember following some is better than none. Even then, you could use this list to make your own to-do list for a hired hand. The fact is, these lists are very easy and a smart way to make your landscapes, lawns and gardens look their best with little effort.

JANUARY

Plant tulip and Hyacinth bulbs along with practically any other bulb that should have been planted in December (or even earlier).

Plant a tree on Arbor Day, usually the third Friday in January! They can be planted along the gulf coast, pretty much year-round. See the Trees Chapter in this book for a more essential list.

Prune established trees, especially those that haven't been touched in years, and do it while they are in their highest state of dormancy.

This the start of the "Fruit Tree Sale Season." at County Extension offices, and even at specialty nurseries that now make it a goal to carry fruit trees year-round.

Start preparing soil areas/beds/raised beds with the garden soils you want for vegetable gardening. It allows the soil, amending with a healthy amount of compost (see the Soils Chapter in this book) to rest and/or mellow before you plant a seed or a transplant as early as mid-February.

Check evergreens, like hollies, hawthorns, Japanese blueberries etc. for scale insect. Tiny little white dots up and down and on the underside of leaves. Kill them with Malathion before they start leaving a lot of sooty mold. Organically, this is the month to suffocate them into death with a Dormant Oil spray.

Feed cool season annuals (ex: pansies, snapdragons, cyclamen etc.) one more application of slow-release blooming plant food. This is the last application before we think about changing color in March.

Take in mowers, edgers, chainsaws, hedge-trimmers, blowers any power tool, battery operated or fuel-based, and have them sharpened and tuned up. When was the last time you had your mower blade sharpened? It should be done at least twice a year.

Incorporate organic matter into existing vegetable gardens. Put out 2 to 3 inches of compost and till it thoroughly.

FEBRUARY

Apply Pre-Emergent herbicides, especially the 2-in-1 types like Barricade or Dimension, to block the pesky weeds of early spring (both grassy and broadleaf weeds). See my fertilization schedule in the Lawn Care Chapter of this book for more details.

This is the time for the "Early Green Up" per MY FERTILIZATION SCHEDULE. Usually this is done with what we call a fast-acting 15-5-10, versus a slow-release fertilizer during the spring and summer fertilization.

Start the pruning process of fruit and nut trees. We mostly encourage taking out dead, damaged or interior branches that may never get sun again. We encourage light pruning and shape-up pruning to the entire fruit or nut tree at the end of the month. We never encourage 'wholesale' pruning.

Time to prune the Crape Myrtles. Valentine's Day is always the best day. Just don't commit Crape Murder. Only prune the tips and scraggly growth. We do not participate in the Annual Crape Myrtle Massacre, where people take the limbs back to nubs each year.

Do not prune small, ornamental trees such as Dogwoods or Oriental Magnolias.

Keep using "cool season" herbicides, like Bonide Weed Beater Ultra and Fertilome Weed Free Zone to control broadleaf weeds like clover, while the temperatures still allow. (See Herbicide section of the Lawn Care chapter if you need clarification) Add surfactant as needed and instructed.

Prune back Hybrid Tea, Floribunda and Shrub Roses in mid-February. Again, Valentine's Day is always a good reference date. Do not prune climbing roses until they have done their bloom cycle.

Pinch back or dead-head cool season annuals, such as pansies or cyclamen, to give them a final push until it's time to change out the annual color come March.

Apply Dormant Oil Spray to any fruit tree that is leafless. Apply the same to citrus trees, only if the temperatures are such that no new growth is starting to emerge.

Plant all your late-blooming bulbs now: Amaryllis, Cannas, Gladiolus etc.

Great time to prune groundcovers, so they will spread liberally at the first sign of spring.

Use soil activators in the lawns and landscapes to super charge nature's own microbial activity in the soils.

Azaleas will lose a small percentage of leaves, as they focus energy on the soon-to-burst-open blooms. Don't panic and think you have to spray a fungicide or insecticide.

MARCH

Feed Azalea and Camellia plants that have finished their blooming cycle and mark your calendars for another feeding six weeks later. Bag your grass clippings at least one time to remove material that could be future thatch.

Buy your slow-release lawn food, and be prepared for the April fertilization. (See Lawn Care Schedule for more details.)
Prune back any azalea or camellia that has done its blooming. Don't wait until new growth starts again, or you'll be pruning off next season's bloom wood.

Start feeding your roses once a month through September, or at the very least once every other month for up to four feedings.

Consider lowering your mower a notch or two early in March to gather up and cut out all the dead or thatch-like grass that browned up through the winter. Then, raise it right back up for the next mowing.

Rebuild landscapes with new shrubs and perennials.

Great time to replenish or introduce mulch. If you have to ask what kind, read the Mulch Chapter in this book.

Prune back established perennials when you see the first sign of new growth.

If you know you have yearly caterpillars or worms that do damage in your lawn or landscape, treat against them with liquid Bt Organic Insect Controls. Bascillius Thurengensis is what Bt stands for.

If March is wet, keep an eye out for slugs and snails. The best control is a two-step approach with the baits around the soil and DE (Diatomaceous Earth) ringed around the base of each plant.

APRIL

This month is the ultimate time for the slow-controlled release part of MY FERTILIZATION SCHEDULE, with the controlled-release types of fertilizers such as Nitro Phos Super Turf 19-4-10. We also call them the 3-1-2 or 4-1-2 ratio fertilizers.

Although bluebonnets and other wildflowers look great in April, they are best seeded in October.

It may be too late to introduce smaller tomato transplants. Rather look for some more mature (in size) looking tomato transplants that are in, at least, one gallon sized containers.

Do. Not. Ever. Use. Weed-n-Feed formulas with Atrazine, if you follow my advice or listen to my program.

When selecting annual transplants, look for short compact plants.

Start feeding Hibiscus and other tropical blooming plants with specific foods (see the "Focused Fertilizers" chapter for details) and do so on an every other month basis.

The best month to repot overgrown/root-bound house plants, or even permanent containerized plants on the patio.

Great month to plant Caladium bulbs.

Yellowing leaves on new growth, that also have dark green veins, is a sign of Iron Chlorosis. Combat this with an Iron and Soil Acidifier combo. Plant roots won't pick up the iron unless the soil is acidified as well.

Time to stock up on perennials, and get them planted. (See the Texas Tough Perennials chapter).

Keep your mower on its highest setting in April and until the fall when we may need to gather or mulch mow leaves.

Time to feed "tropical" flowering plants for the first time, and specifically with foods designed for Hibiscus, Alamanda, Trumpet Flower, and Bougainvillea as examples. Never feed them "bloom booster" fertilizers with very high middle numbers. (See the Focused Fertilizers chapter).

MAY

Continue to fertilize roses. They can be fed monthly or every other month at the very least.

Keep an eye out for powdery mildew on Crape Myrtles and Roses. The easiest method is still a soaking of Consan 20 (formerly Consan Triple Action 20) because you can use it as often as needed.

Start planting the summer annuals like Vincas (a.k.a. Periwinkles), because the soil is warm enough.

If you didn't apply your spring lawn fertilizer in April, get it done now! This is the month for the 2nd application of the 2-in-1 pre-emergent herbicides, per MY LAWN FERTILIZATION SCHEDULE.

Nutgrass and Virginia Buttonweed can and should be controlled with specific herbicides. (see the weed control section of my Lawn Care Chapter for details)

Remember Mother's Day! In fact, don't just buy flowers, buy her flowering plants that will last a whole lot longer than a week.

JUNE

Develop good irrigation/watering practices by June for your lawn. In fact, most grasses only need about 1-2 inches of water per week from Mother Nature and your own irrigation. When temperatures start to hit the higher 90s, we should make further adjustments. (See the Lawn Care Chapter for tips on how to determine that 1-2 inches gets measured out).

If you don't already have some of those tropical blooming plants, noted in last month's to do-list, start planting them in June.

This is the best month to make critical notes, about what needs more water in the heat or what needs more trimming/pruning and what responds to fertilizers best. In other words, keep a gardening journal.

Spider Mites can become a big problem the hotter it gets, especially in the Juniper, Cypress or Arborvitae families. They are best controlled with a systemic insect control, usually containing Acephate. There are many miticides on the market, but they have to be used almost weekly to control this pesky insect.

The best time to be watering your turfgrass is early in the mornings, between 6 a.m and 9 a.m. During the summer months avoid watering between noon and 7 p.m.

If well-watered lawns are suffering from a yellow look, it might need an iron and soil acidifier. Just adding iron may not work unless the soil has the right acidity.

Keep an eye out for Gray Leaf Spot too, which also can cause a yellowing look to lawns, with splotches of brown on each leaf blade. Apply Daconil/Chlorothalonil-based fungicides at 4 ounces per gallon of water in a pump up sprayer, two times over 60 days.

Be on the lookout for spotting on evergreen shrubs like Photinias, Ligustrums and Hawthornes, and head off fungal leaf spot with copper-based or systemic fungicides with PPZ.

JULY

Summer color, like we think of with annuals, doesn't just have to be in flowering form. Consider Caladiums, Coleus and Copper Plants.

Look for Chinch Bugs (see Insects chapter for more detail), which can decimate a St. Augustine lawn in under two weeks if not controlled early. Three applications of liquid insecticides like Bifenthrin over a two week period should work.

First week of July (even end of June) is considered our Summer Fertilization for MY LAWN FERTILIZATION SCHEDULE.

If you want a fall vegetable garden, consider building the beds now. You can start planting things in August.

DO NOT prune trees from now until the first cold front hopefully by October. Despite advertisements you'll hear on every manner of radio call-in shows, summer is the WORST time to prune hardwood trees.

Prune off expired or weak-looking bloom heads of Crape Myrtles if you want any chance at a second set of blooms.

Black Sooty Mold means you have an insect problem such as scale, aphids or whiteflies on all manner of shrubs and trees in the landscape. Solve the insect problem first, then get rid of the BSM with soapy water or Consan 20. (See the insect chapter in this book, if you need more detail)

Evergreen plants with berries, such as hollies, will show you they are suffering from drought stress by dropping berries. Maintain consistent irrigation practices.

How is your watering/irrigation regimen? Remember, that as temperatures increase, so should your irrigation regimen.

AUGUST

Begin establishment of your own compost pile. When leaves fall, yards are mowed and old layers of mulch need to be trashed, compost them instead. Plus, you will be ready for all the autumn leaves to build and even better, richer compost. Ever wondered why they call the best Leaf Mold Compost?

Dead head (pinch off expired flowers and seed heads) on just about every summer annual and perennial if you want more productive growth and more flowers through September.

Another great month to be out in front of specific weed control like Doveweed, Virginia Buttonweed and Nutgrass.

Continue to keep an eye out for fungal leaf spot on evergreen shrubs like Photinias, Ligustrums and Hawthorns. Spray systemic fungicides or Copper-based fungicides again.

It's not too late to introduce heat-loving annuals/bulbs like Caladiums, Coleus, Copper Plants, Vincas, Zinnias and Marigolds.

They will require extra water early on after transplant, but they should thrive right up to the first frost.

Some trees that are under heavy stress will shed lots of leaves in August. Don't panic though, because it's usually just a defense mechanism. Trees are dead or dying, if the leaves turn brown and don't fall off the tree.

Fall Webworms – the actual webs in the leaves and small limbs of trees – can and should be controlled with any liquid insecticide. For future prevention, spray liquid Bt on all green leaves over a two month cycle.

Scan for Brownpatch, the circular looking fungal disease predominant in St. Augustine lawns. For organic-minded gardeners, treat with leaf mold compost or with bio-inoculants such as Micro Gro from MicroLife.

SEPTEMBER

Big month for planting easy-to-grow vegetables for fall gardens: Tomatoes, peppers and squash. Don't wait until month-end, no matter how hot you think it might be.

Watch for grub worms. If you see more than 4 or 5 per square foot area, then you should treat with Dylox. Most grub worm controls from 20 years ago don't exist, and the ones with Imidicloprid don't control the kind of grub worms we have.

Keep an eye out for Brownpatch - the fungal disease again. This disease loves areas that have too much moisture from irrigation systems.

Despite the show of cool season annuals such as Pansies and Snapdragons at mass merchandisers and big box stores with their garden centers, please refrain from planting until cooler temperatures, hopefully by mid-October.

Transplants (people not plants) need to know that September does not mean cooler weather along the Gulf Coast. If you remember that, it will save you a lot of aggravation.

Many nurseries and garden centers start having clearance sales. Many times they promote upwards of 75% off. This is actually a good time to stock up on bargain plants, considering that October and November are two of the best months to do transplant work.

If any plant, such as Crape Myrtles, Roses or Hibiscus are covered with aphids, just blast them off every morning with a strong stream of water. It's the smartest, most environmentally safe way of controlling insects during any month, but especially prior to cooler weather when plants want to go dormant.

If the weather changes and high temperatures are below 80, it's time to consider the "cool season" herbicide applications for control of pesky problems like Doveweed and Virginia Buttonweed.

OCTOBER

October and November are considered two of the best months to transplant trees and shrubs; or frankly redo landscapes. It's enough time to establish roots before winter, and actually the cheapest rates for full service landscape companies.

It's finally the perfect month to prune Oleanders.

October is also considered one of the best months to transplant any containerized tree, no matter the size of the root ball.

This is the month for what we consider the 2nd pruning season for roses, especially shrub roses like Knock Outs, as well as Hybrid Tea, and Floribunda.

This is the optimum month to start planting all manner of spring flowering bulbs, with the exception of the ones that need to be chilled for 6-8 weeks.

This is the first opportunity we get to winterize/fall fertilizer our yards, per MY FERTILIZATION SCHEDULE. This is a fall feeding, and should not be considered a way to green up the the yard even more.

Scan for fungal diseases in the yard, like Brownpatch, if the temperatures are still very warm, especially during a spread of 80 degree highs and 60 degree lows.

We don't cover a lot of veggie gardening in this book, but 10-15 is considered the optimum time to plant 10-15 Sweet Onions – hence their name.

October is THE month to separate existing bulbs and perennials. Divide them and get them in new places in the landscape during this month.

This is also the month to STOP feeding many things. No more fertilizer for citrus, trees, shrubs, roses etc. Give them a chance to start the dormancy process.

NOVEMBER

Apply a new layer of mulch, or even compost as mulch, to protect for upcoming winter months and let it serve as a barrier against weeds.

If you haven't done your 'winterizer' or 'fall-feeding' of the lawn, do so now.

Be on the lookout once more for scale insects. On any evergreen shrub, Malathion is still the best scale control, however, if the weather is cold enough, apply dormant oil spray.

Plant your cool season annuals, like Pansies and Snapdragons etc., if the temperatures weren't cool enough in October, they are now.

Cyclamen are considered the ultimate cool season annual for shade. Always wait until Halloween to plant them. Which is why November is always the best month moving forward.

Fallen leaves and pine needles make instant mulch for many flower beds. There are some leaves that contain tannins and do not make a good mulch. Avoid leaves from Pecans, Hickories and Live Oaks.

Remember to vote! Usually the first Tuesday in November. (That's the most political this book, and my radio show will ever get.)

If you have a fall vegetable garden, consider picking off green tomatoes if a freeze is coming. Fried green tomatoes and homemade chow-chow are much better alternatives to freeze-damaged fruit.

Thanksgiving is considered the best time to plant bulbs like narcissus and daffodils.

Prepare to move and prep Plumeria for winter storage and forced dormancy.

DECEMBER

If you know a freeze is on the way, water all in-ground plants and even potted plants thoroughly. A dry root zone is more susceptible to freeze damage

Give gardening gifts to the 'green thumbs' in your family during the holiday season. pH Meters, gardening books (such as this one), hand trowel kits etc. all make good, reasonably-priced gifts.

Find the nearest Christmas tree farm www.texaschristmastrees.com to choose and cut your own, super fresh Christmas tree.

There is little rain in this month, which means you still have to augment Mother Nature and keep some moisture out there once every other week, if we aren't getting rains.

If you haven't applied a layer of mulch in November, do so NOW!

You can still be doing major pruning jobs on hardwood trees. They are in their highest state of dormancy between now and January.

Remember there are several other cool season annual choices other than pansies and snapdragons. Calendula, Sweet Alyssum, Stock, Dianthus and Ornamental Kale and Cabbage. And another reminder that Cyclamen are the greatest cool season annual for shady areas.

The 'average' annual first freeze/frost date in this region is mid-December. So, have all your work done – mulching, watering, winterizing the lawn, transplanting and investment in frost covers.

If you've been working a fall vegetable garden, now is the time to get the unusual Cole Crops in like mustard greens, turnips, leeks and even green onions.

If you get Bonsai plants as gifts during the holiday season keep them outdoors to grow and indoors for show. They are actual dwarf versions of trees, so they need Mother Nature's air and sunshine. "In for Show; Out to Grow!"

If you purchased bulbs in October requiring refrigeration, get them into the ground now: Tulips, Hyacinth and Crocus as perfect examples.

Final

A Salute to the Greater Houston Ace Retailer Group (GHARG)

Once again, I have to heap an unbelievable amount of gratitude towards my friends with the Greater Houston Ace Retailer Group. If they weren't so supportive, and such a benefactor towards the publishing of this and my last two books, I don't think my writings make it into near as many hands over the past 10 years.

GardenLine the radio show has had a great relationship with this group of Ace Hardware stores for over a decade now. The local Ace Hardware stores are a great resource for most of the products on my fertilization schedule. And Ace Hardware does it all with great customer service. The kind of customer service that has earned Ace Hardware the J.D. Power & Associates award for best customer service among hardware stores, 12 out of the last 13 years.

I'm not a fan of big box stores and that's no secret on the radio show. Big box stores don't carry those GardenLine Endorsed products, and their customer service is virtually non-existent, specifically in the garden centers. Sure you may think you can get an ant killer, or a weed killing product for a few pennies less at a big box store, but you're not getting anything like the customer service you get at Ace Hardware stores, especially those here in southeast Texas – the majority of which are part of the Greater Houston Ace Retailer Group.

And not to be the ultimate "homer" for these Ace Hardware stores, but they are all privately-owned and operated. And when you support any business like that, you're really helping the local economy even more. So, in a way, my endorsement of the local Ace Hardware stores is not just because of what they carry for GardenLine customers, it's because of all the good they do for their customers, the community and the charities they support.

Their local fundraising efforts for the Children's Miracle Network and the Texas Children's Hospital in our community is phenomenal. One of the ways they do that at virtually every store here in southeast Texas is called "Round Up The Dollar." In my personal opinion, it is one of the smartest fundraising efforts I've ever seen, and yet it is so utterly simple. Once you've had your purchases totaled up at the cash register, they'll ask if you want to "round up your total" to the next dollar. So, if your tab was $16.55, you would be donating .45 cents to make your total 17.00, but knowing that 45 cents would be donated to their fund raising efforts with the Children's Miracle Network. More specifically, it's all staying local by going to the Texas Children's Hospital. Over the past 10 years, they have exceeded donations totaling over $2 million dollars through this and a few other efforts like the $5 buckets.

Just five years ago, the 35 GHARG Ace stores of the Houston area outpaced the 170+ stores of the Chicago area, by a Texas mile when it came to charitable giving. You can always be Texas Proud of your local Ace Hardware stores when it comes to what they are doing in so many charitable ways. And a shout out to how generous Texans can be when it comes to a cause.

One more reason to visit an Ace Hardware by you is specific sales. One in particular is the 5 gallon bucket sale, where everything you can fit inside a Children's Miracle Network Bucket is 20% off at checkout (As noted earlier the bucket is just $5 too, but all of that goes straight to CMN and TCH, and who doesn't need several 5 gallon buckets around the garage?) Like rabid Black Friday shoppers, I look forward to that sale every year. They also do a similar 20% off sale once-a-year, on everything you can get in one of their patented and sturdy, brown paper bags.

At print time, there are currently 35 stores that consider themselves part of the Greater Houston Ace Hardware Group, which reaches from Brenham to Beaumont. But there are plans to have anywhere from 3-5 new stores also open up within the year 2020.

Once again, let me thank so many of the great owners/ operators of these stores that GardenLine has had such a great relationship with, and I look forward to meeting those new owners in the not-too-distant future. For my listeners, and readers of this book, remember that Ace Hardware has been around for a long time, and will be around for a long time to come. And as long as I'm hosting GardenLine, the radio show, I will always want them to be among our partners.

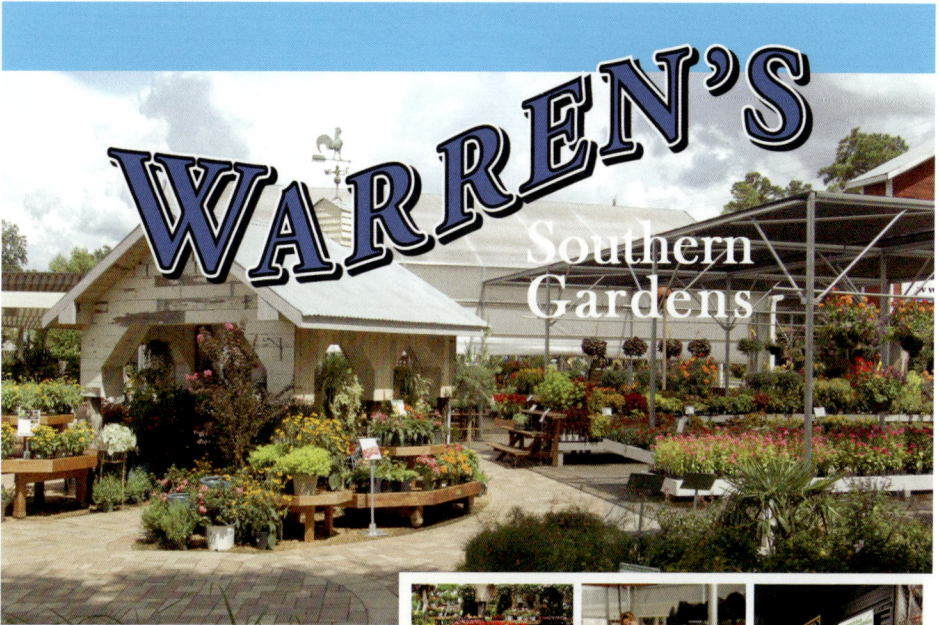

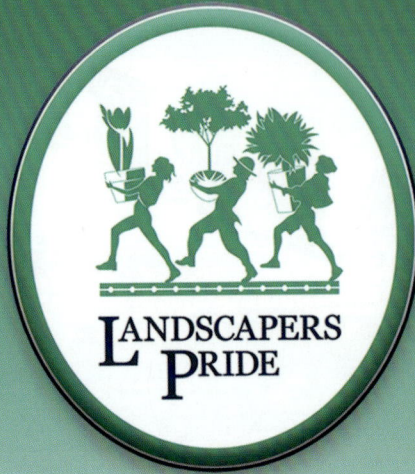

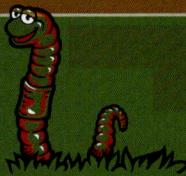

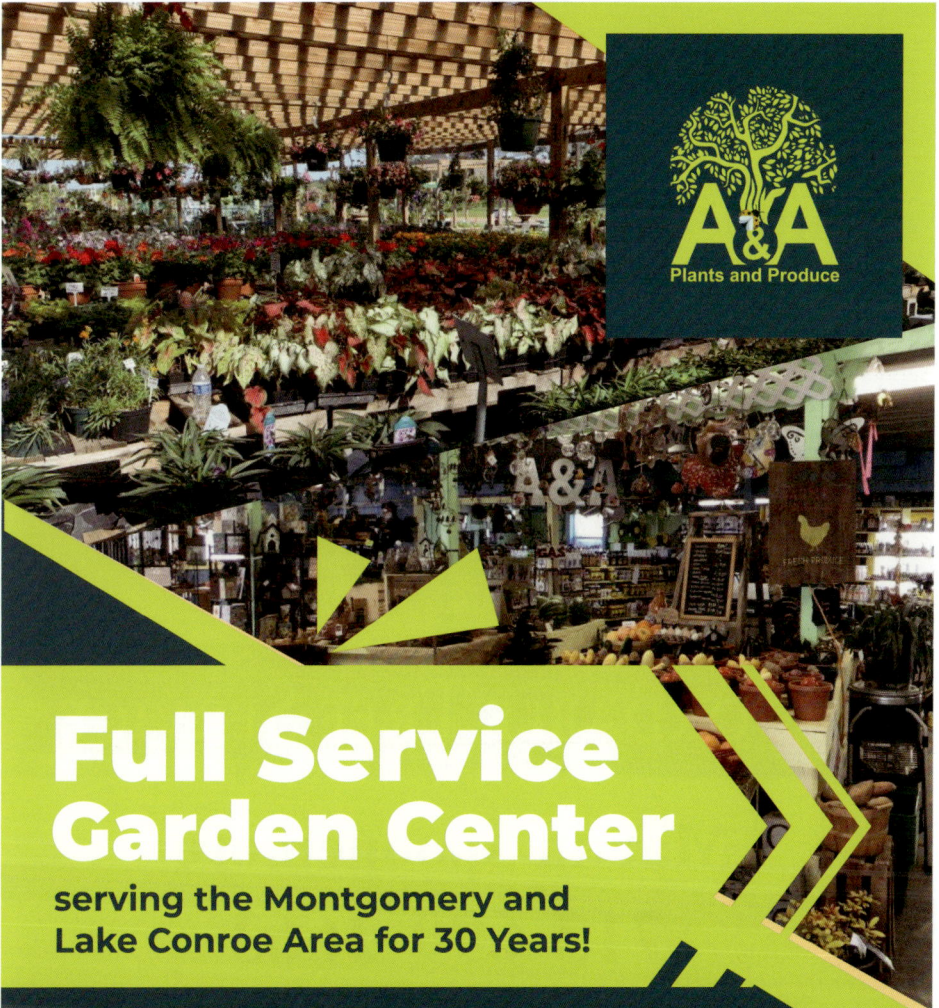

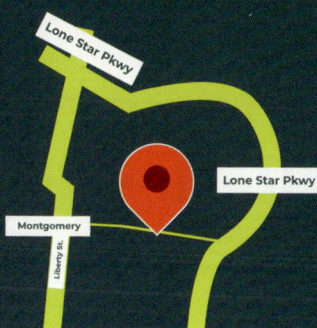